Preschool and Kindergarten Numbers and Counting Practice Workbook 1 - 12

Marina Vigodsky

Copyright © 2018 Marina Vigodsky
All rights reserved.
ISBN: 172746009X
ISBN-13: 978-1727460094

Introduction

The Preschool and Kindergarten Numbers and Counting Practice Workbook 1-12 introduces children to numbers, counting and addition of numbers 1 to 12.

Use of a three-sector introduction button for teaching numbers provides contextual background which makes the notion of quantity and its symbolic representation – an actual number – easier to grasp. From cognitive standpoint common contextual background turns all three parts of a three-sector number introduction button into one unit which contributes to better memorizing of numbers. The second button is a recall and rehearsal button. Being connected with the first one it contributes to efficient learning and recall. Two buttons represent a teaching unit. Both teaching buttons provide cognitive framework for improved introduction, rehearsal and recall of numbers.

Addition exercises in the end of the workbook provide practice for initial manipulations with numbers.

1-12 numbers chart.

1-6 numbers chart.

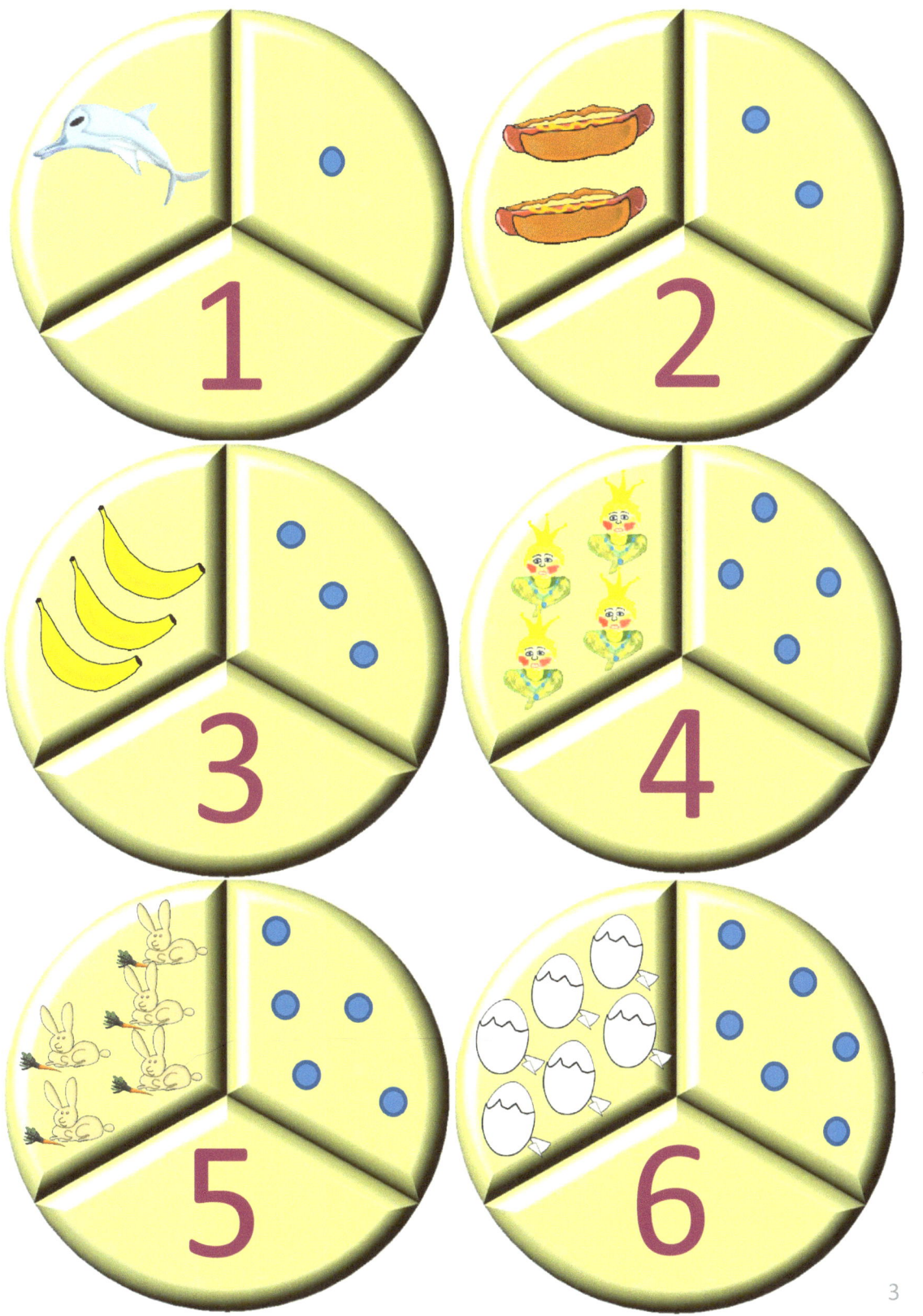

Fill in the missing number like in the big button below:

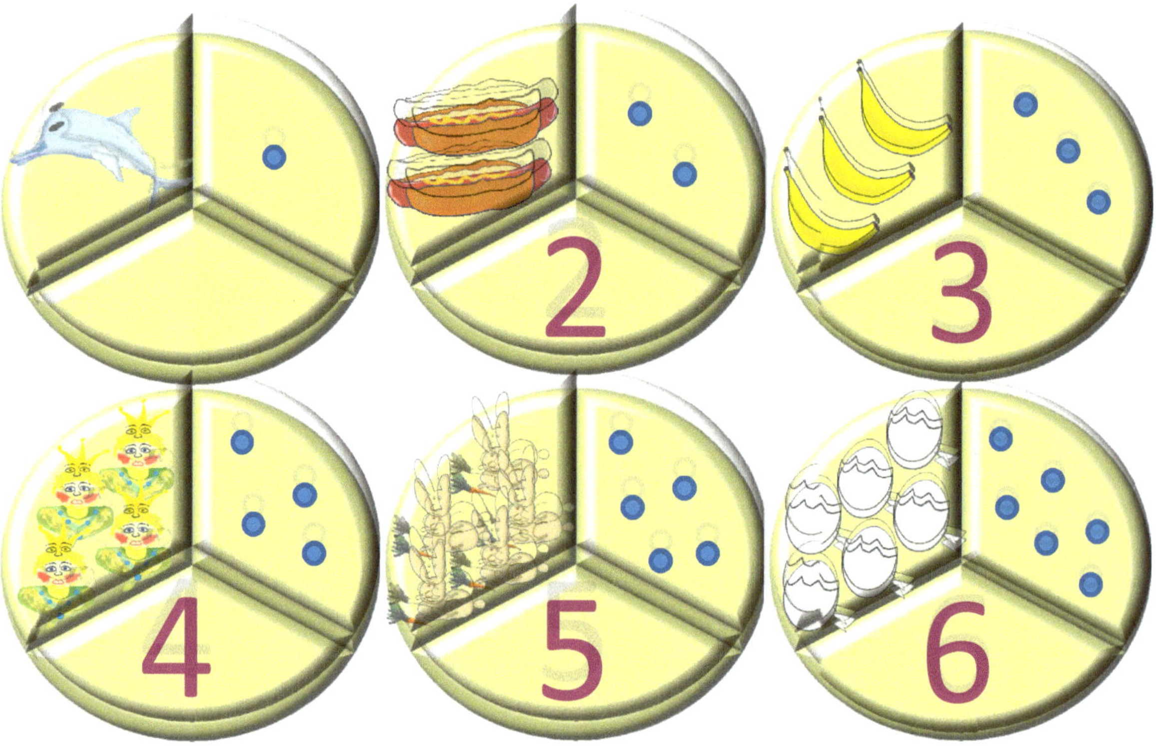

Fill in the missing number like in the big button below:

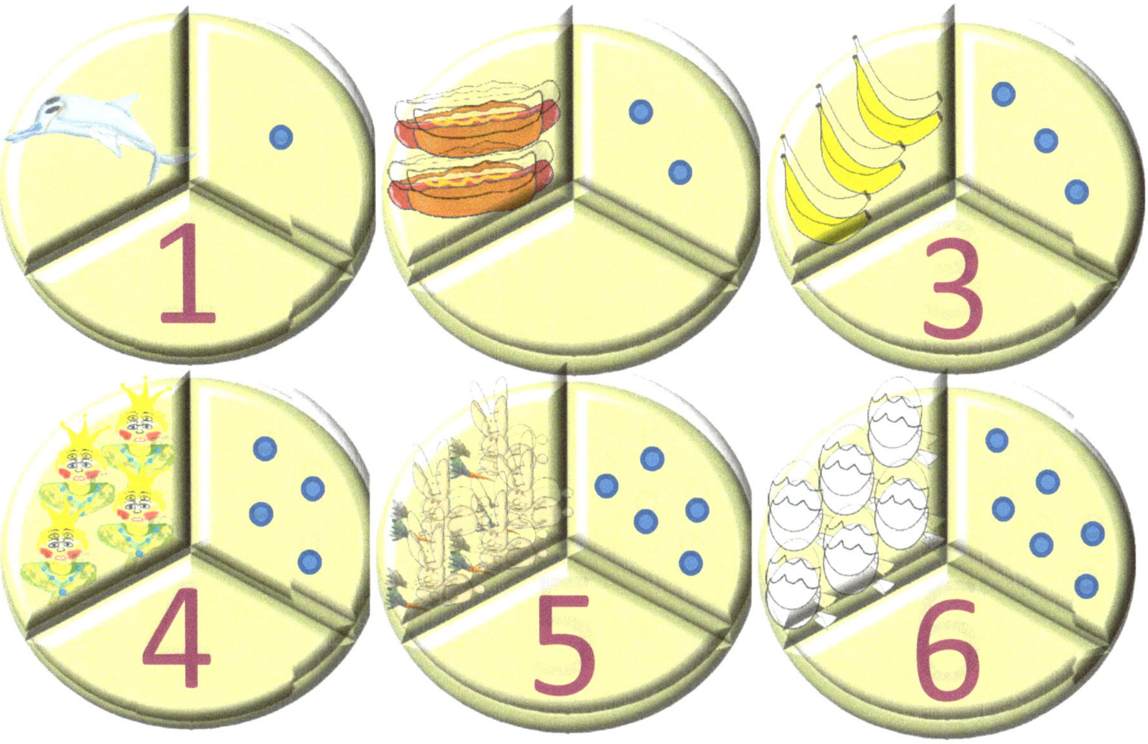

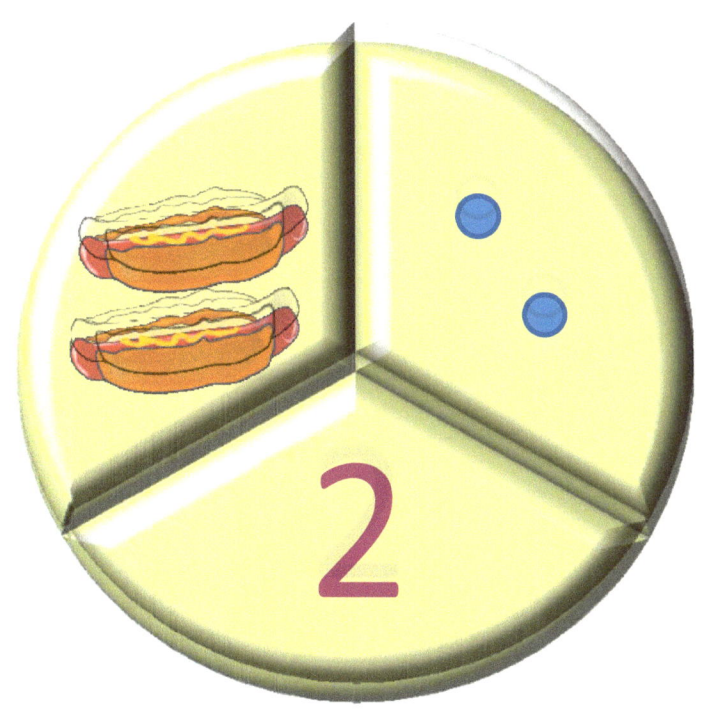

Fill in the missing number like in the big button below.

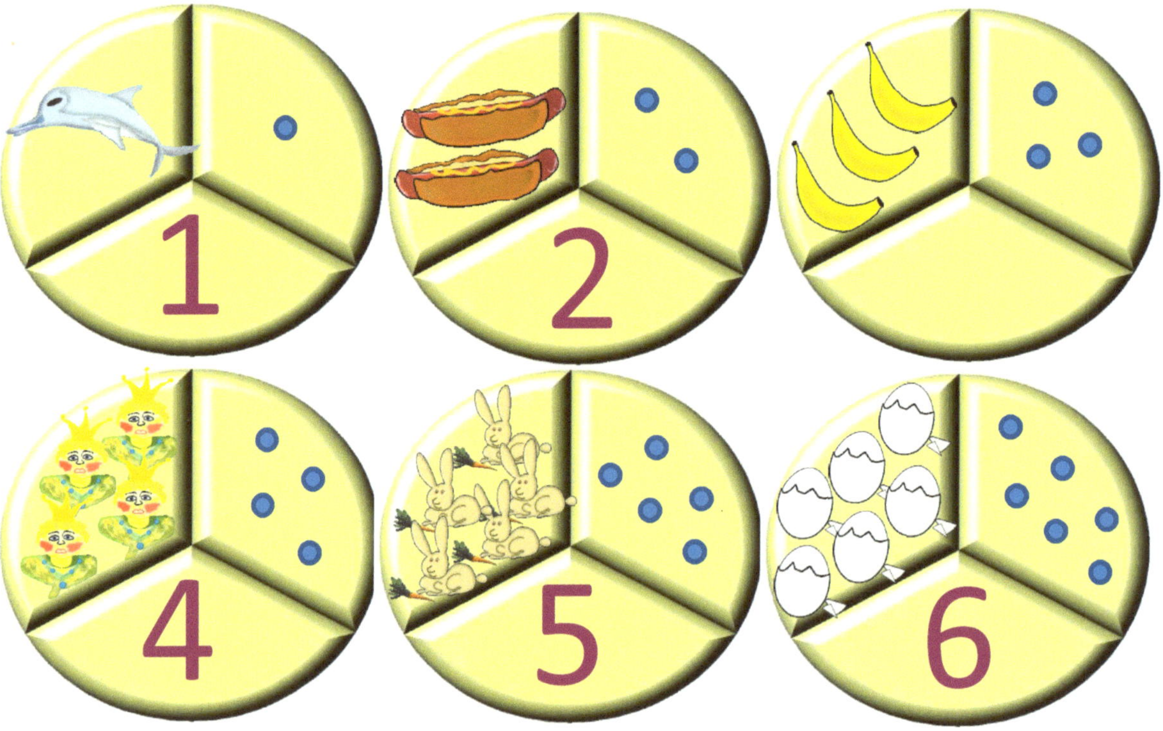

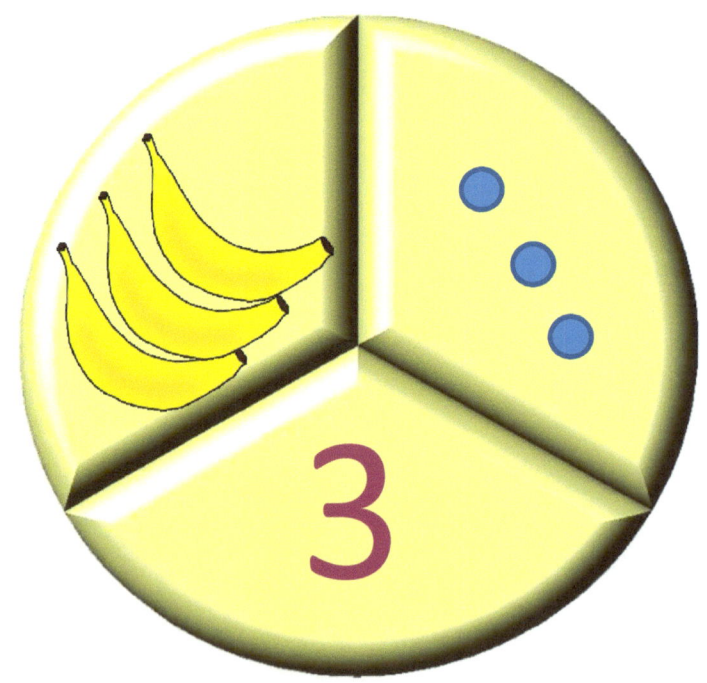

Fill in the missing number like in the big button below.

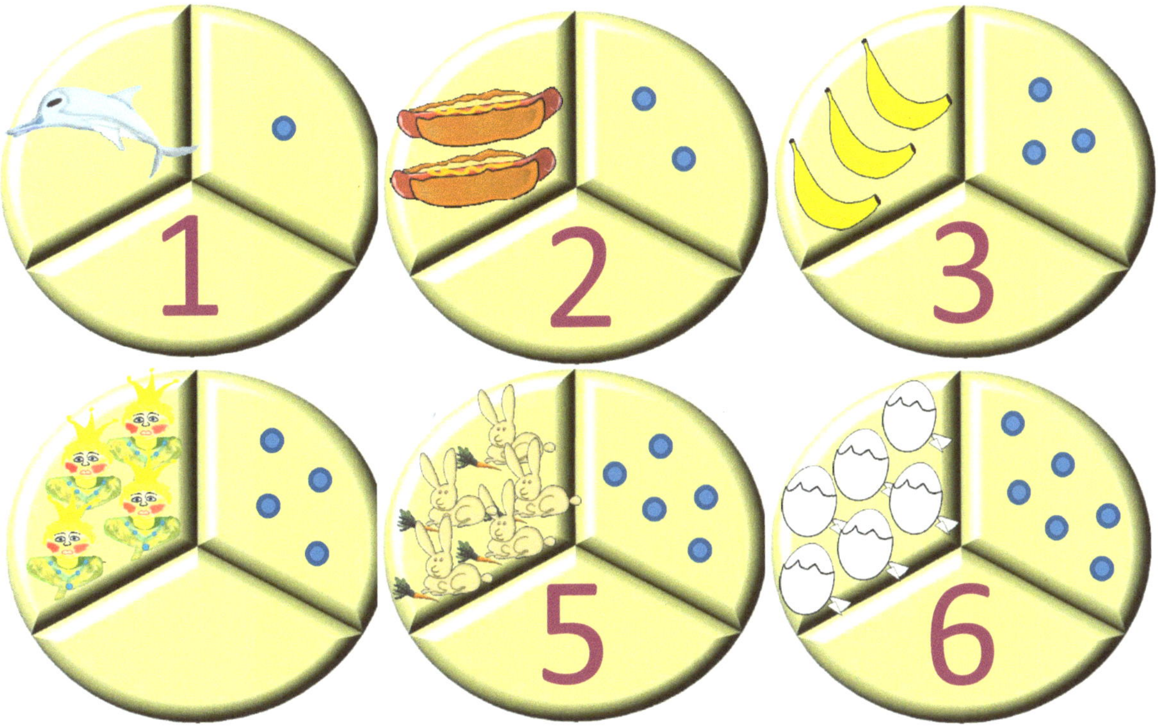

Fill in the missing number like in the big button below:

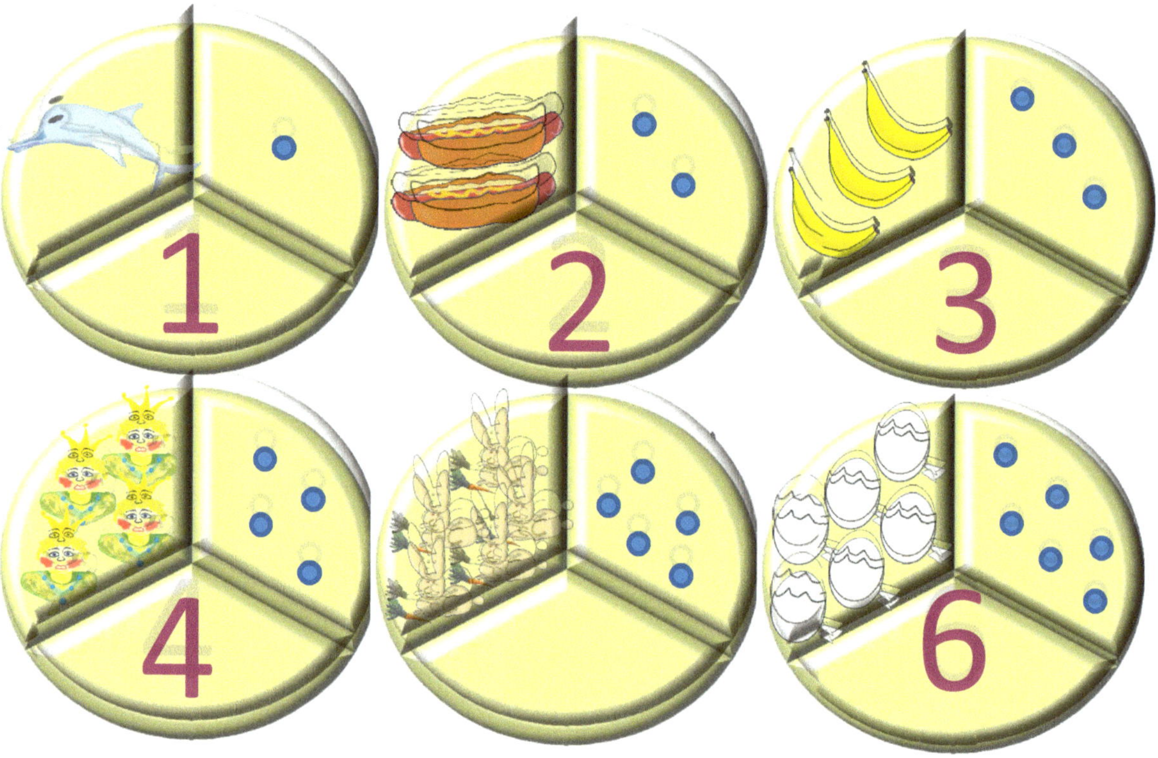

Fill in the missing number like in the big button below:

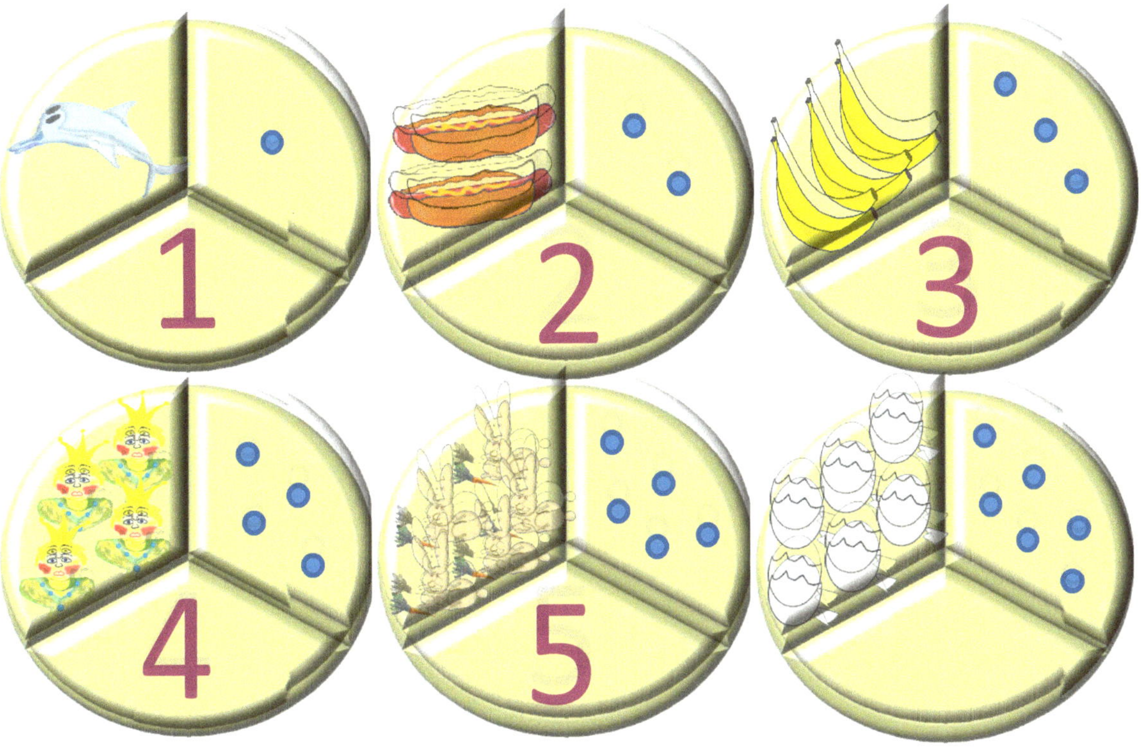

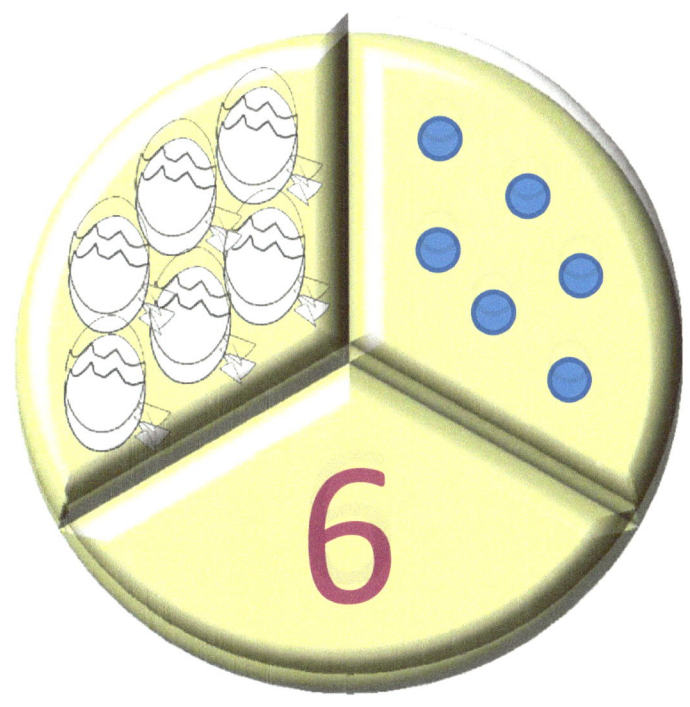

Fill in the missing numbers.

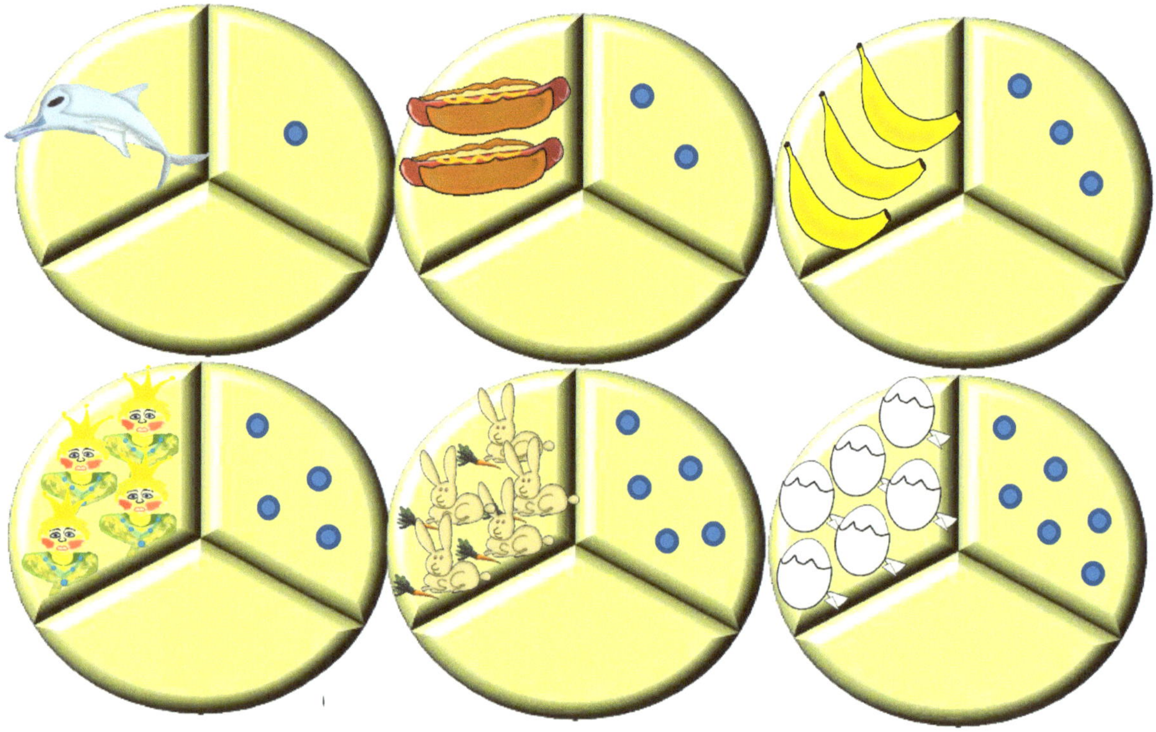

Fill in the missing number.

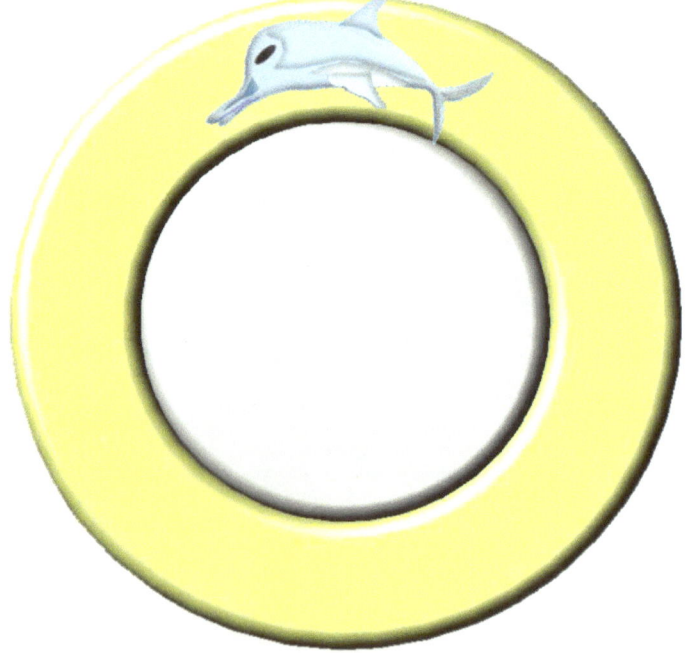

Fill in the missing numbers.

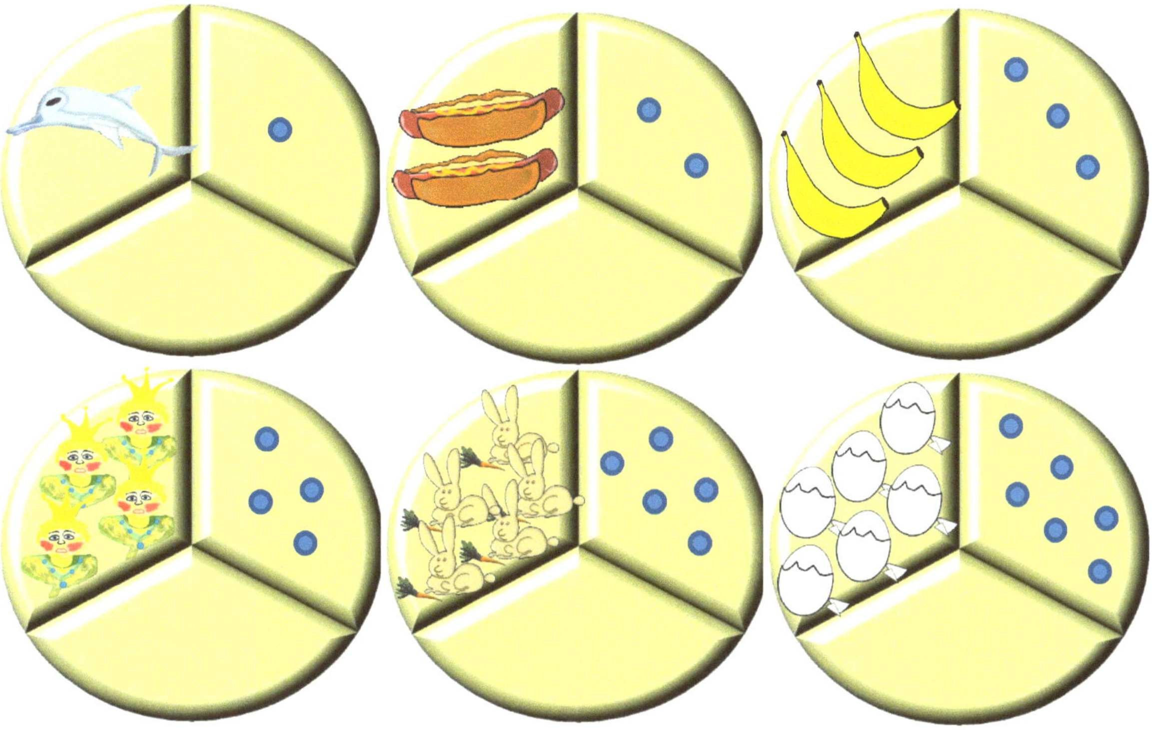

Fill in the missing number.

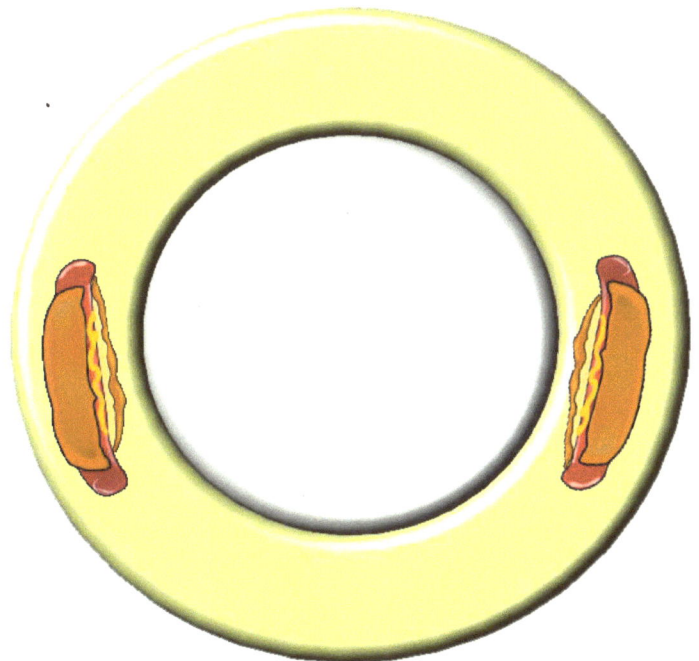

Fill in the missing numbers:

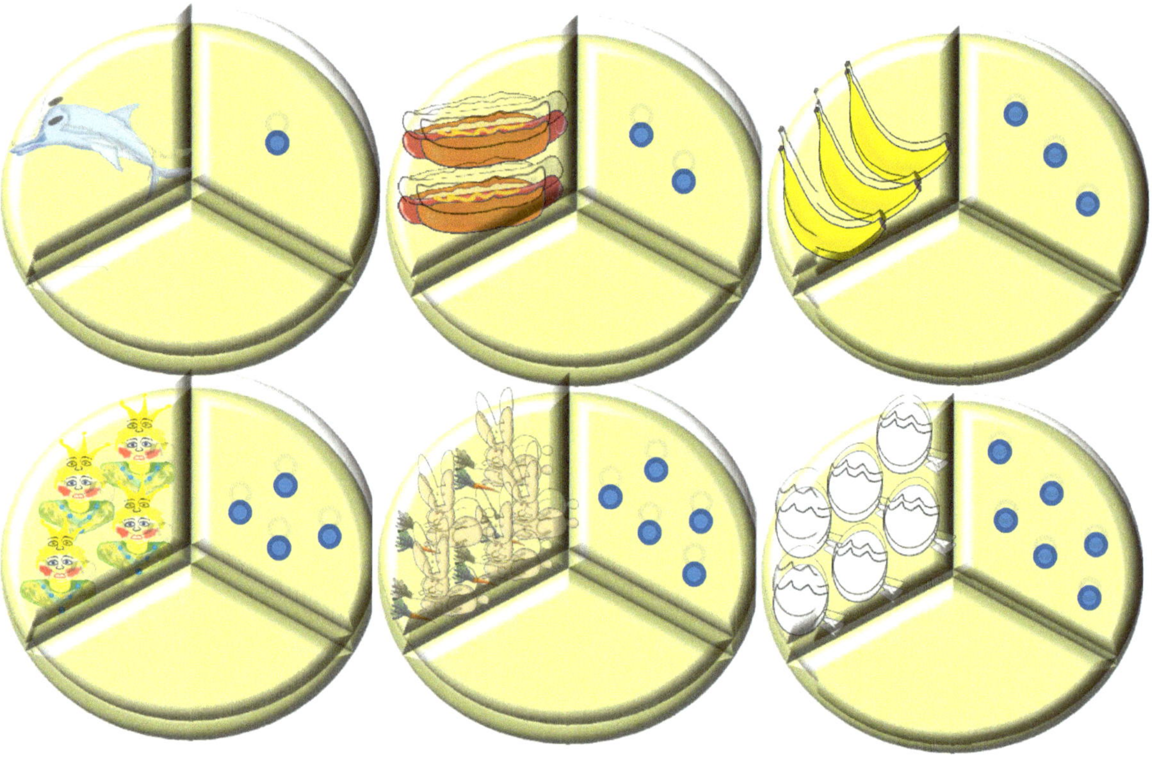

Fill in the missing number.

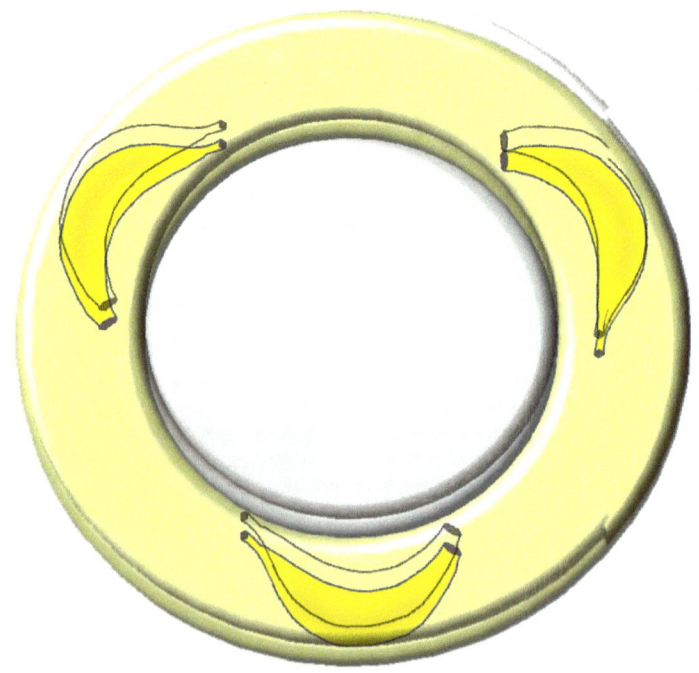

Fill in the missing numbers:

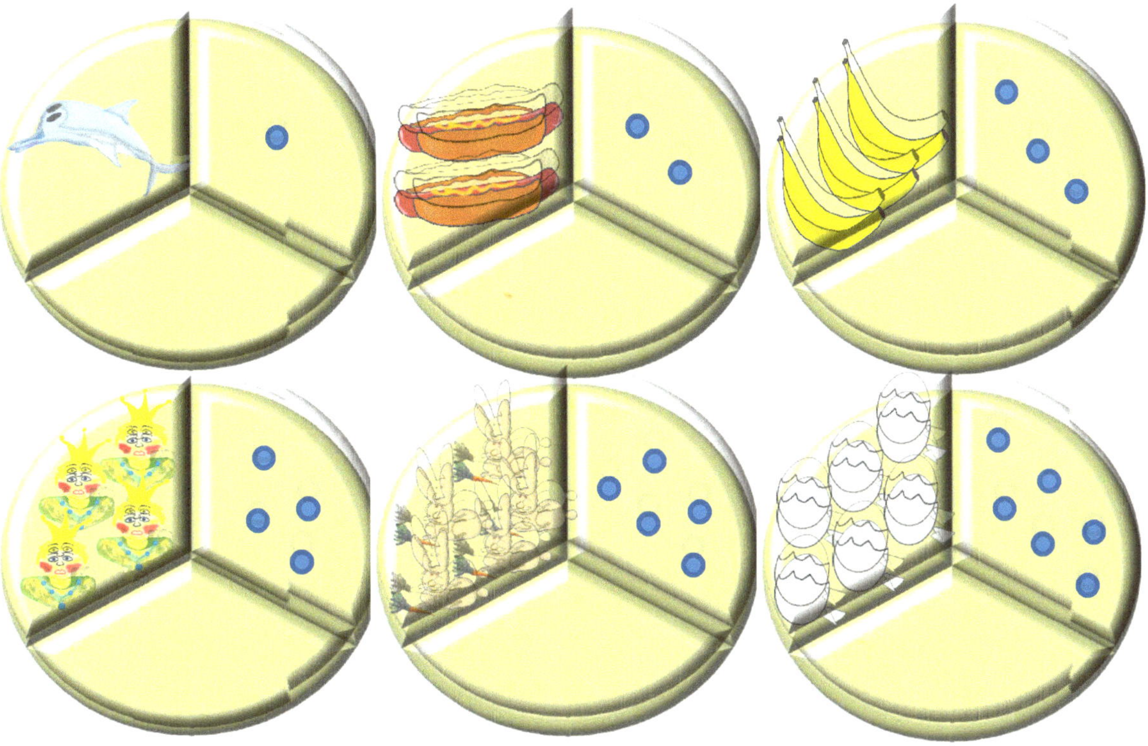

Fill in the missing number.

Fill in the missing numbers.

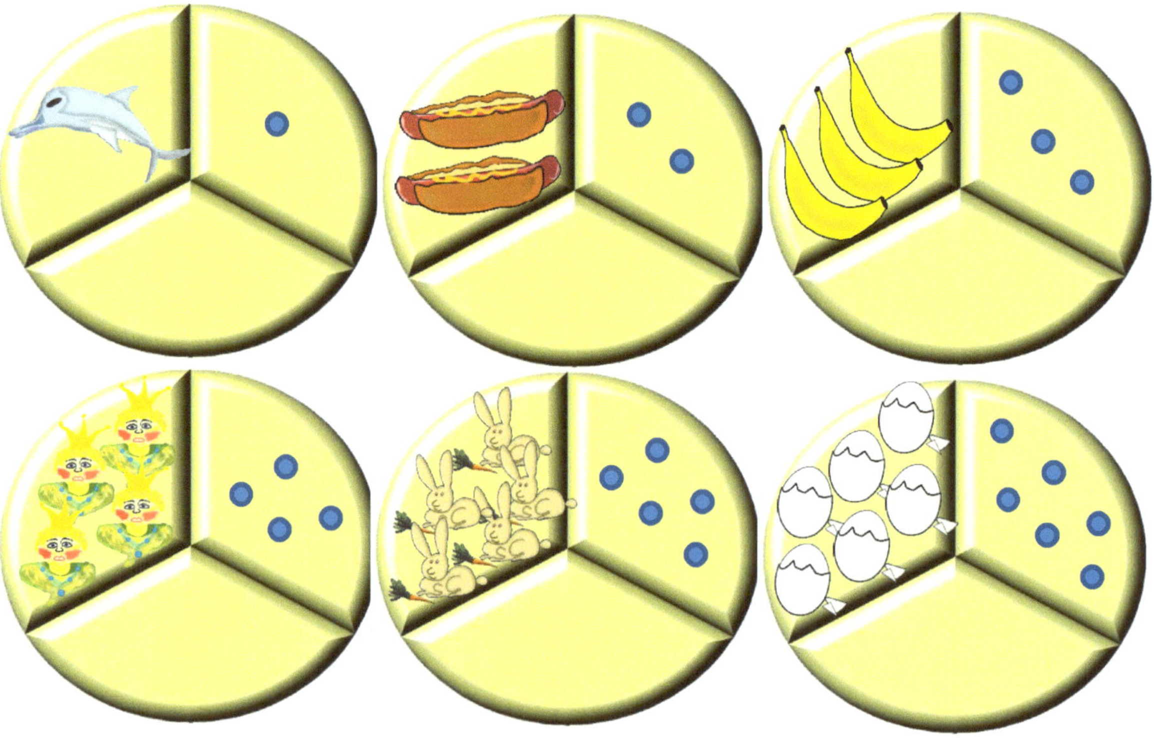

Fill in the missing number.

Fill in the missing numbers.

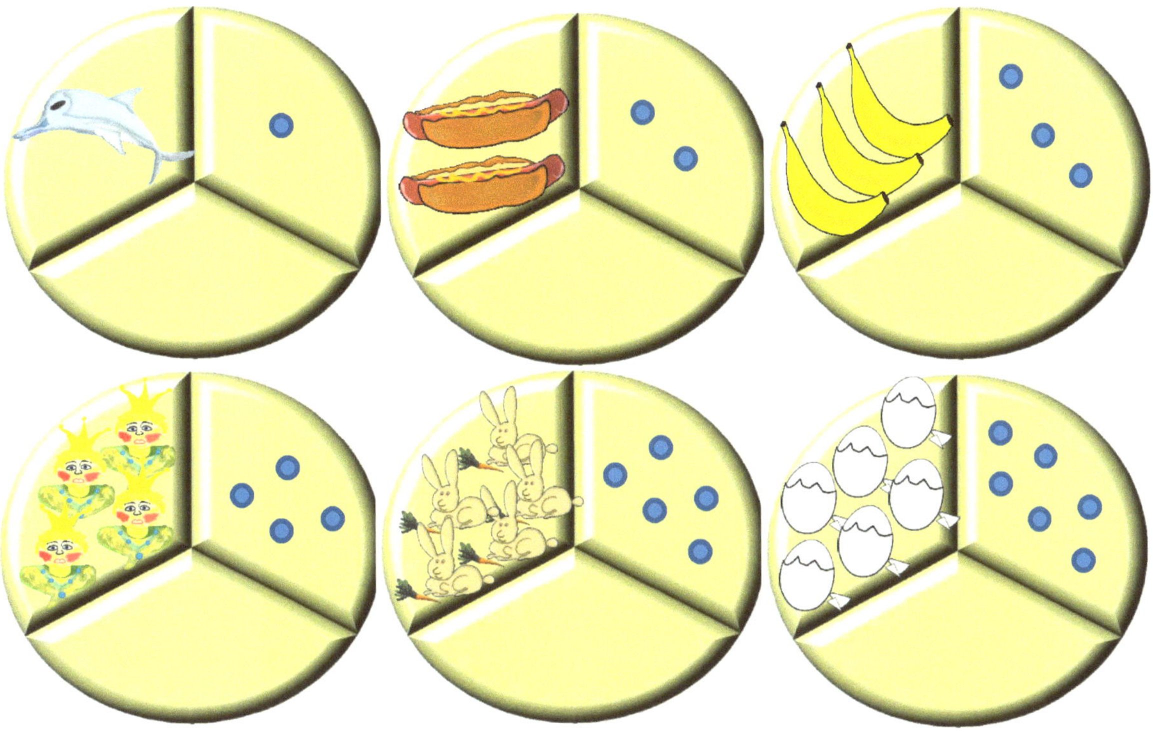

Fill in the missing number.

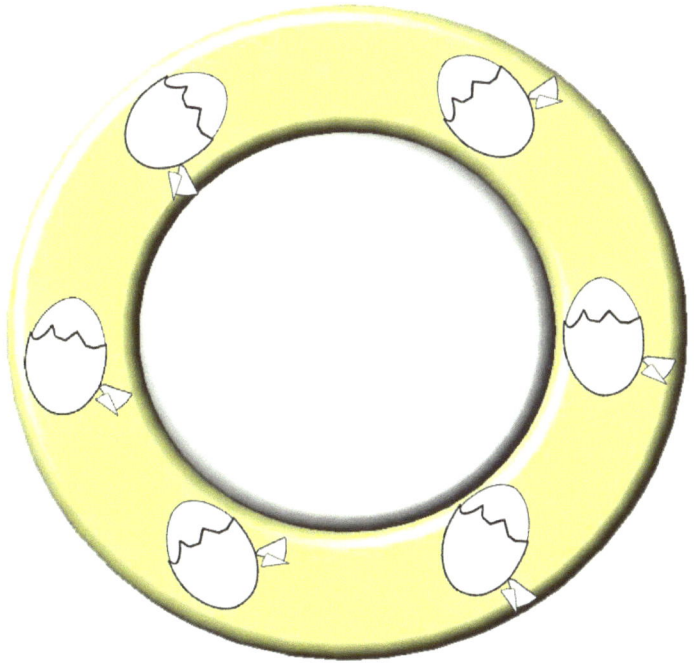

Fill in the missing numbers:

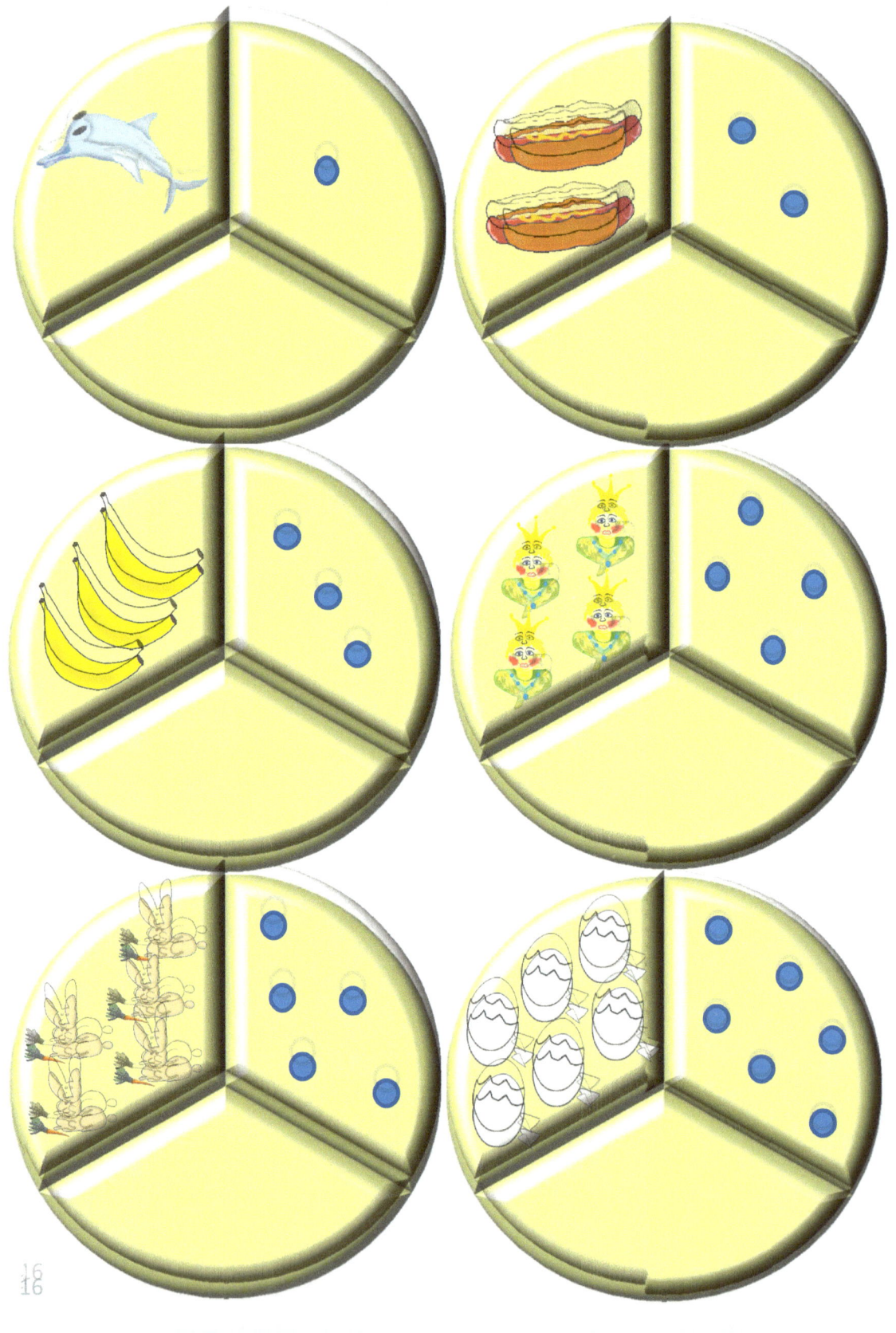

Fill in the missing dots and numbers:

Fill in the missing numbers.

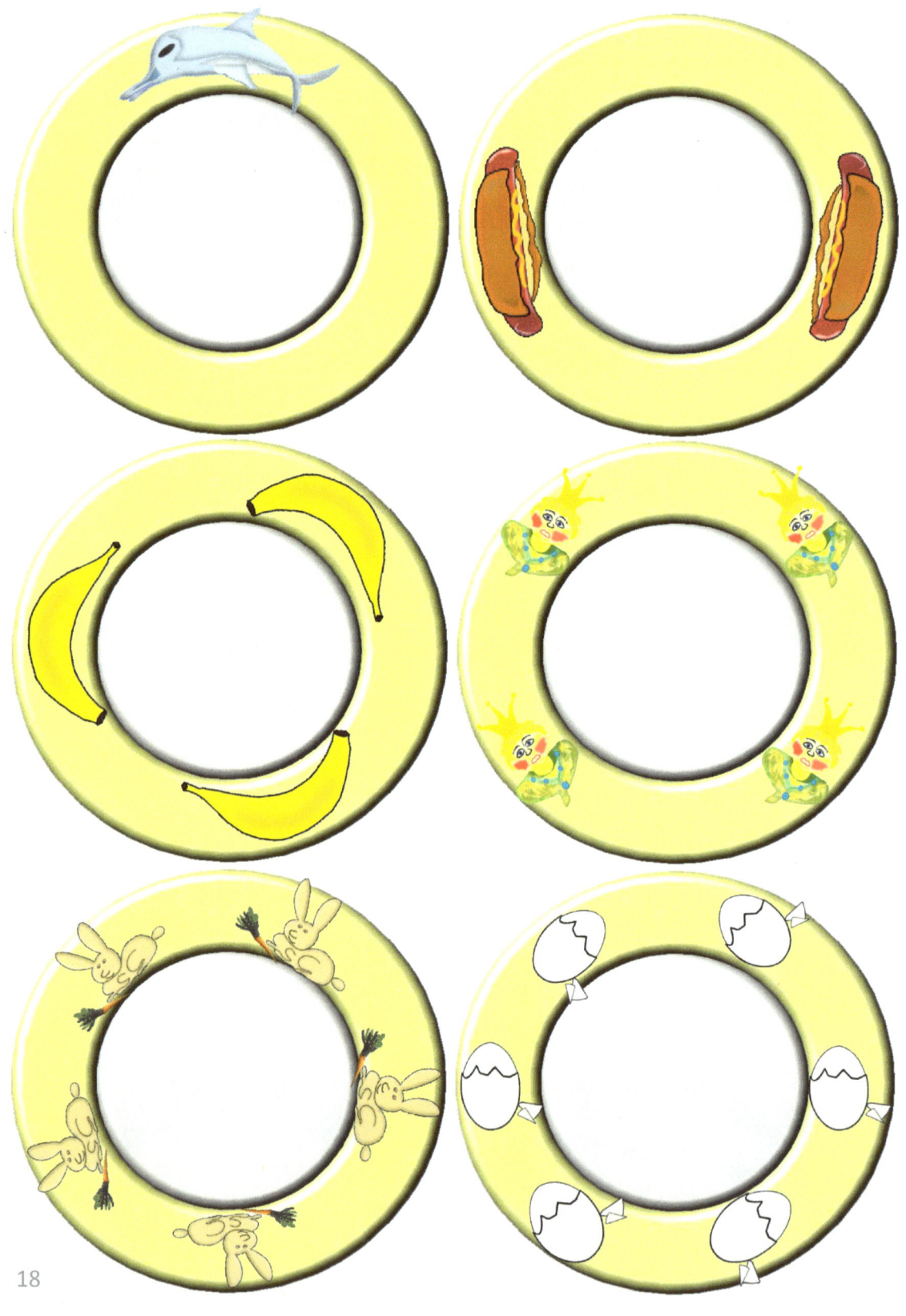

Fill in the missing numbers.

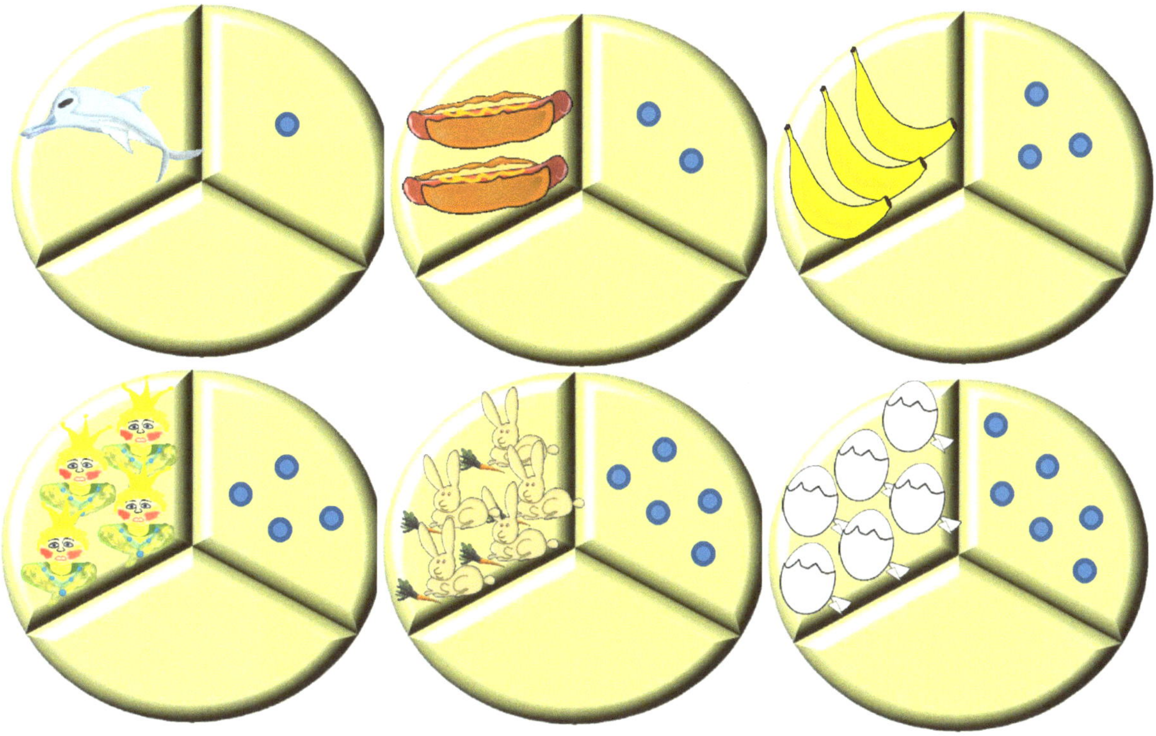

Fill in the missing numbers.

19

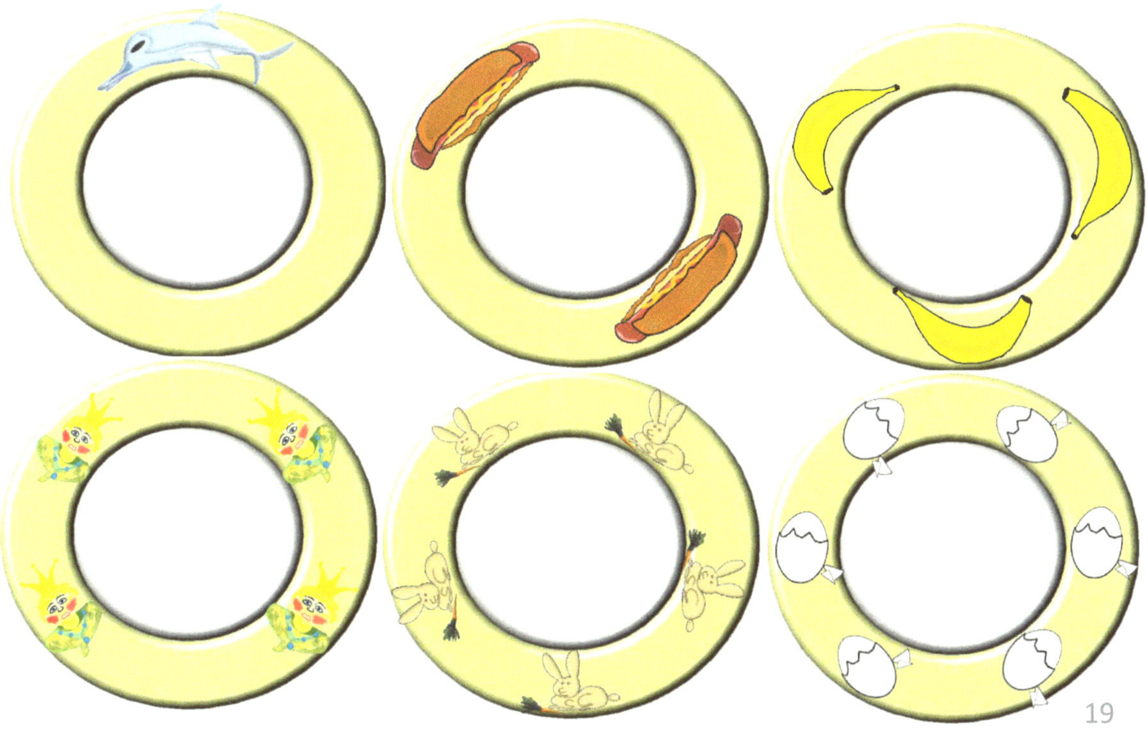

Fill in the missing dots and numbers.

Fill in the missing and numbers.

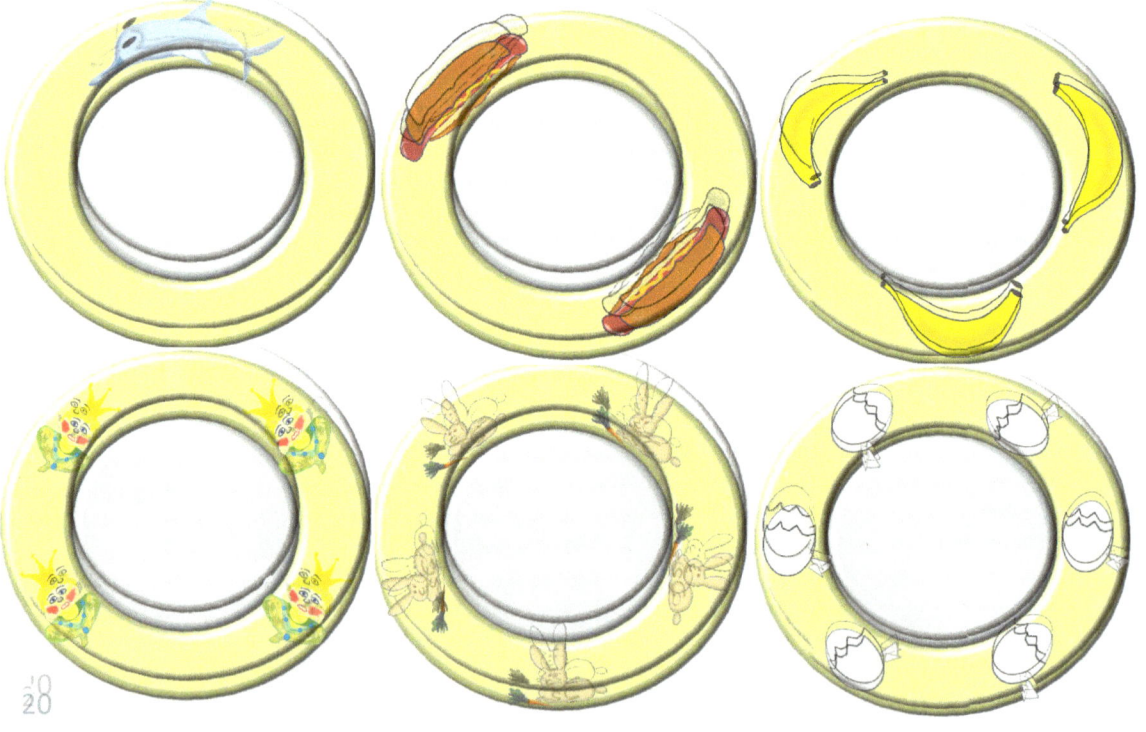

7-12 numbers chart:

Fill in the missing number like in the big button below.

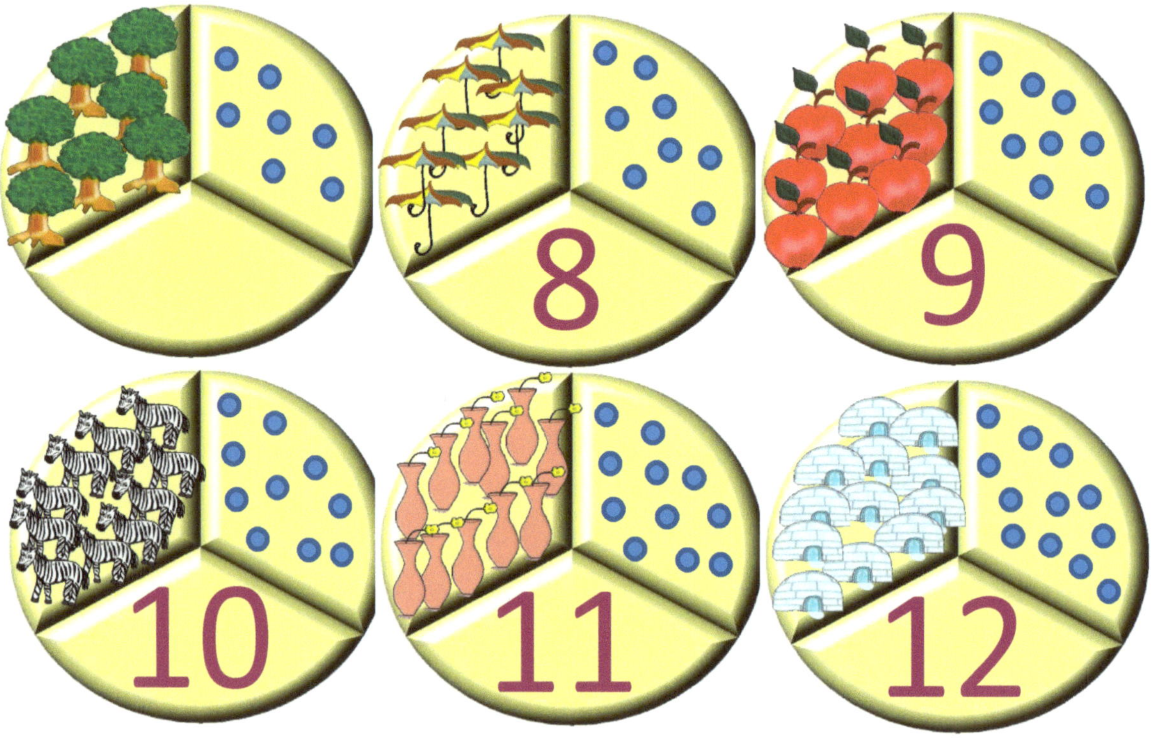

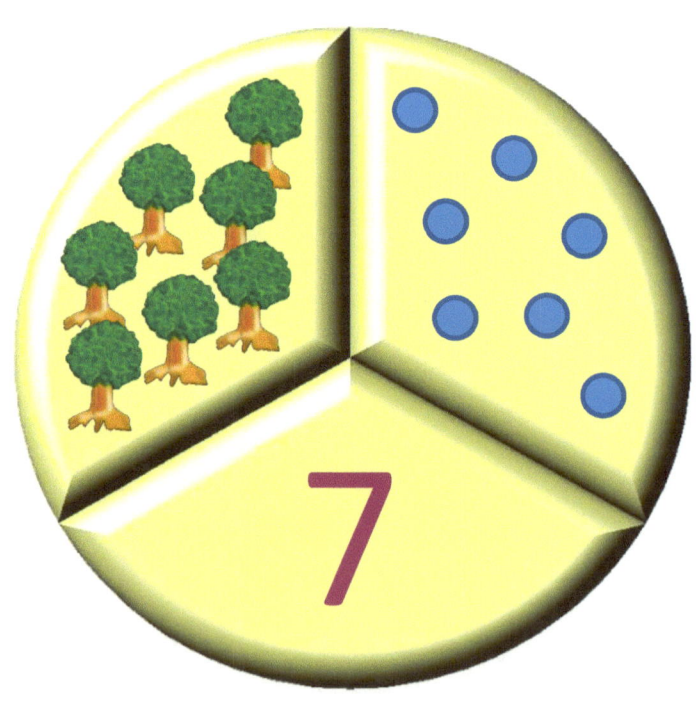

Fill in the missing number like in the big button below.

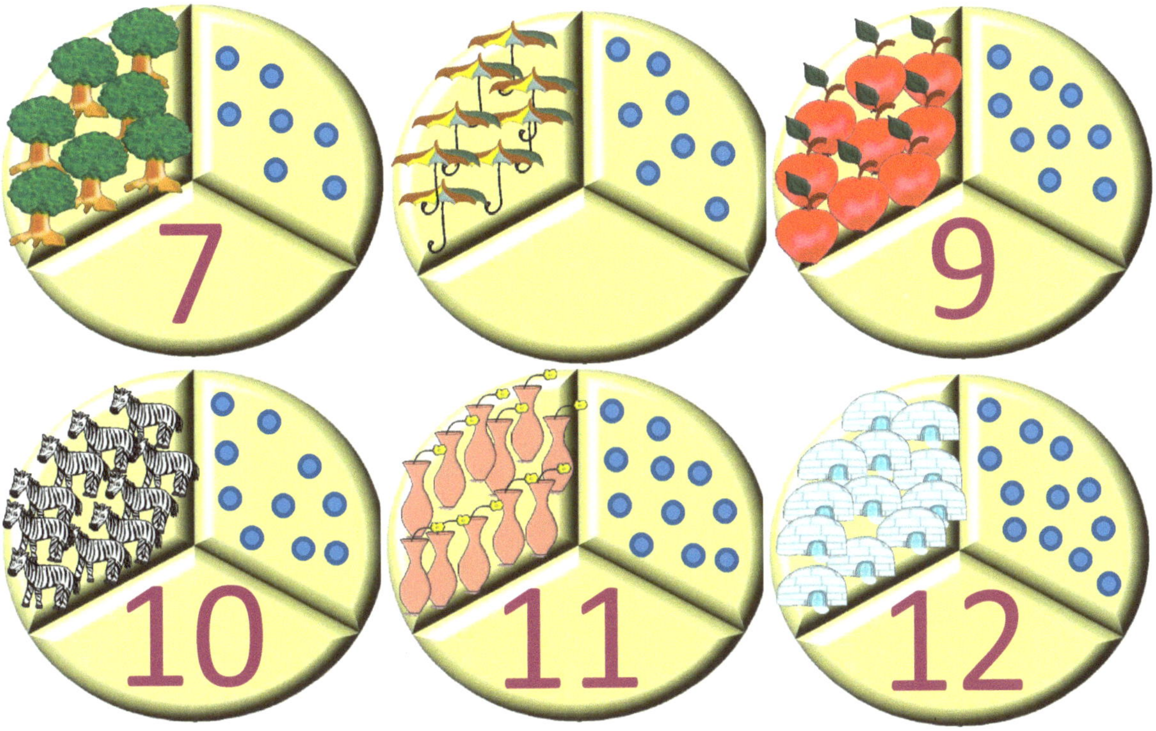

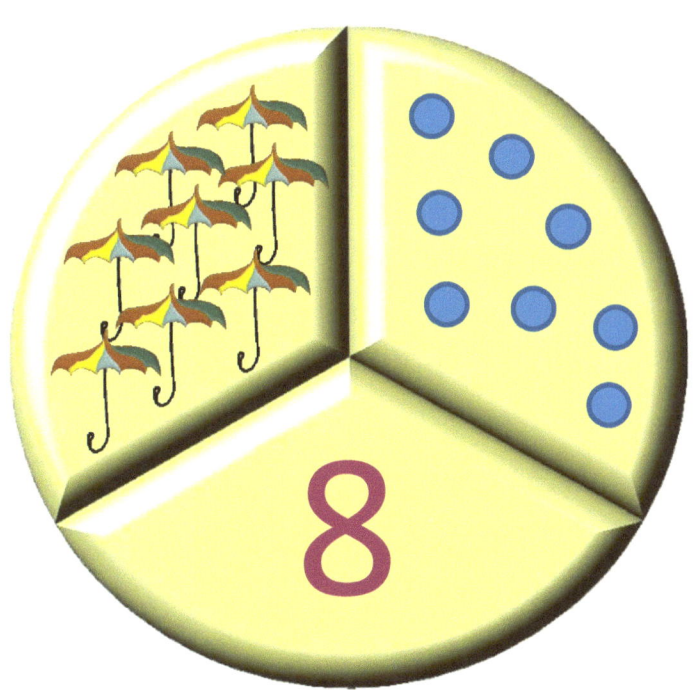

Fill in the missing number like in the big button below:

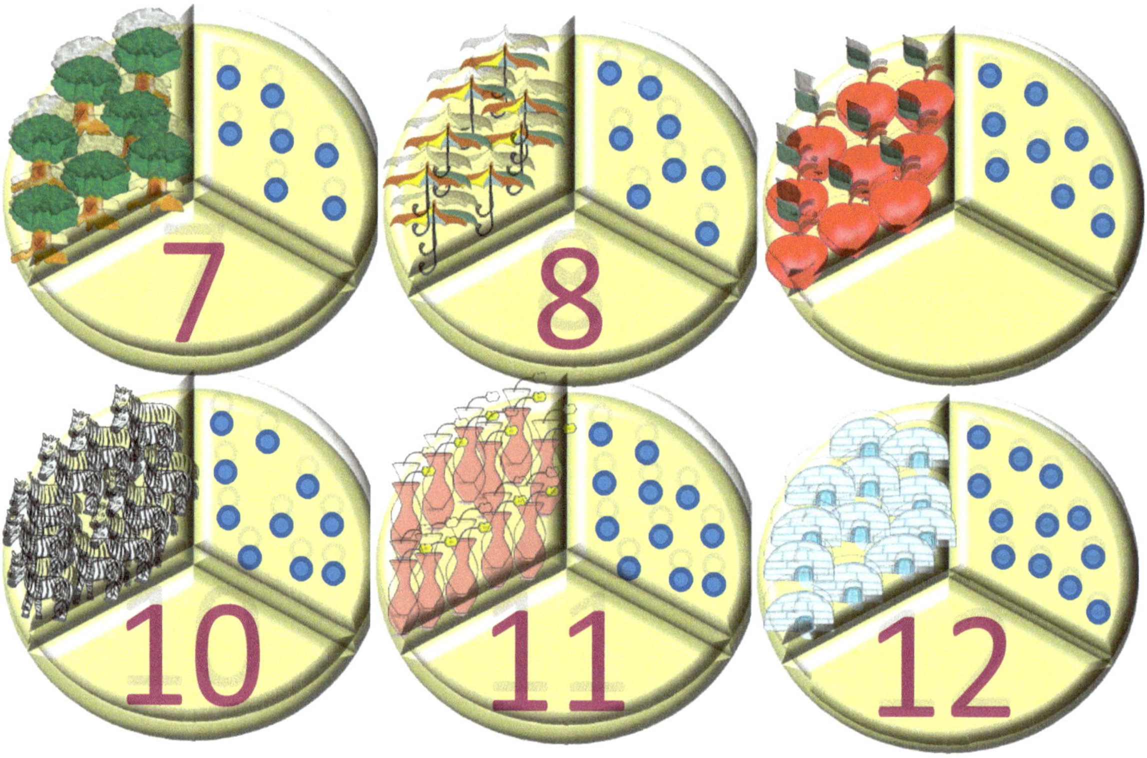

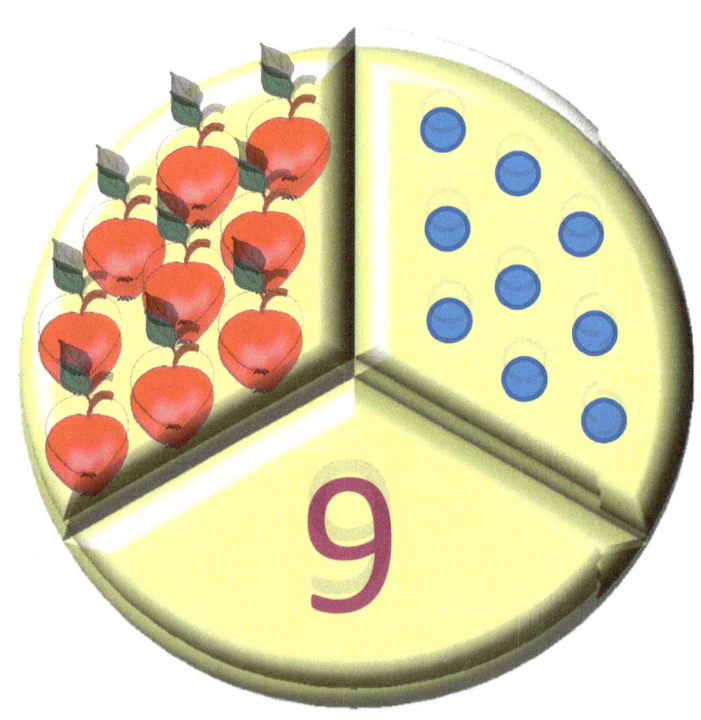

Fill in the missing number like in the big button below:

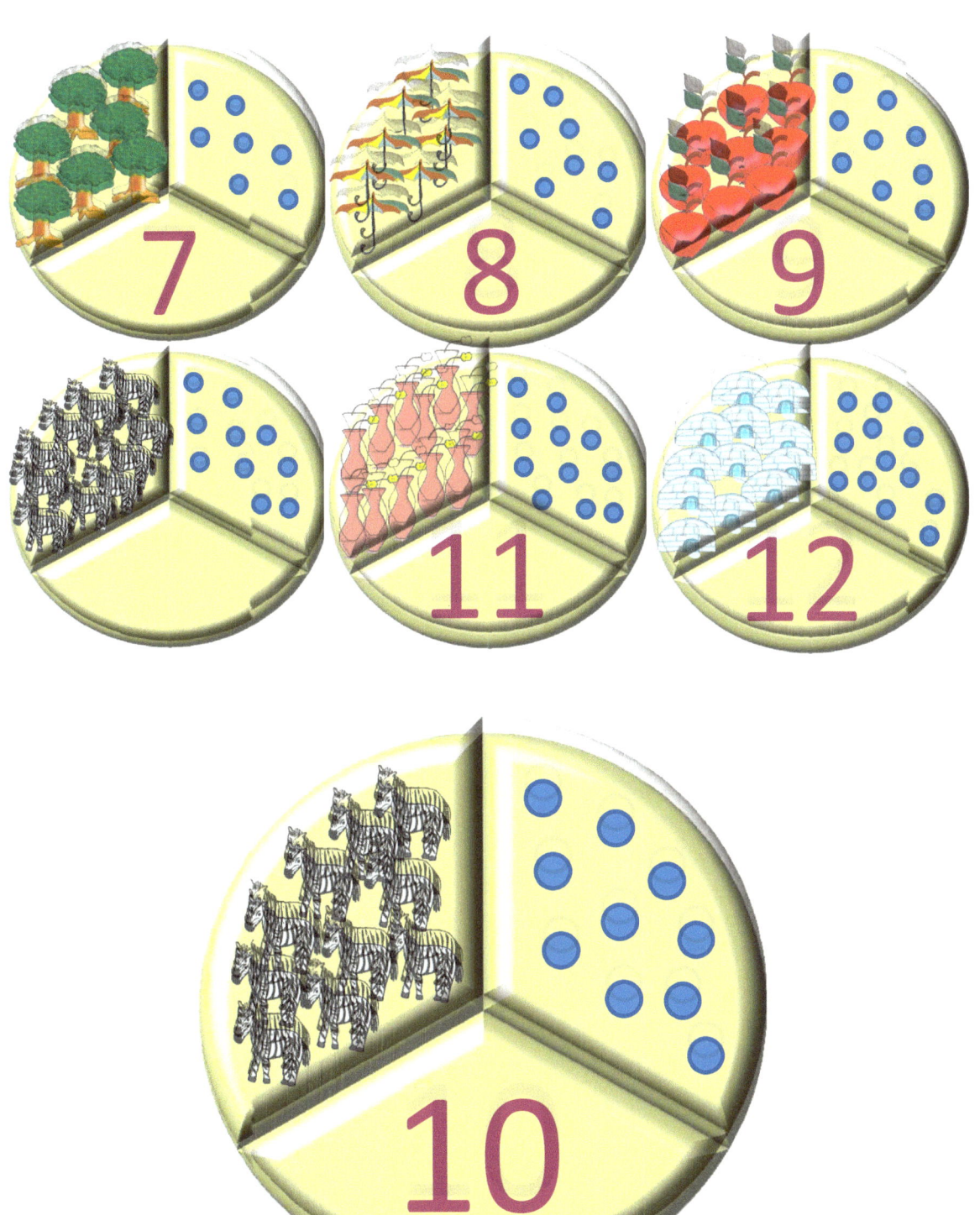

Fill in the missing number like in the big button below.

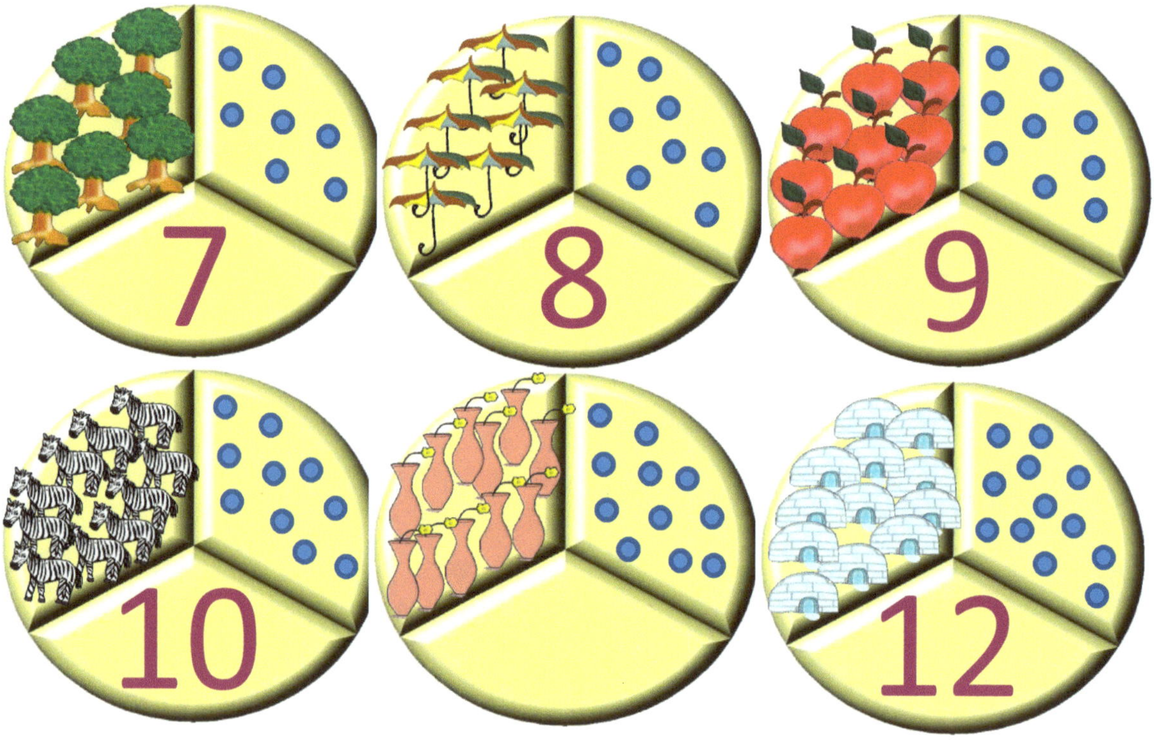

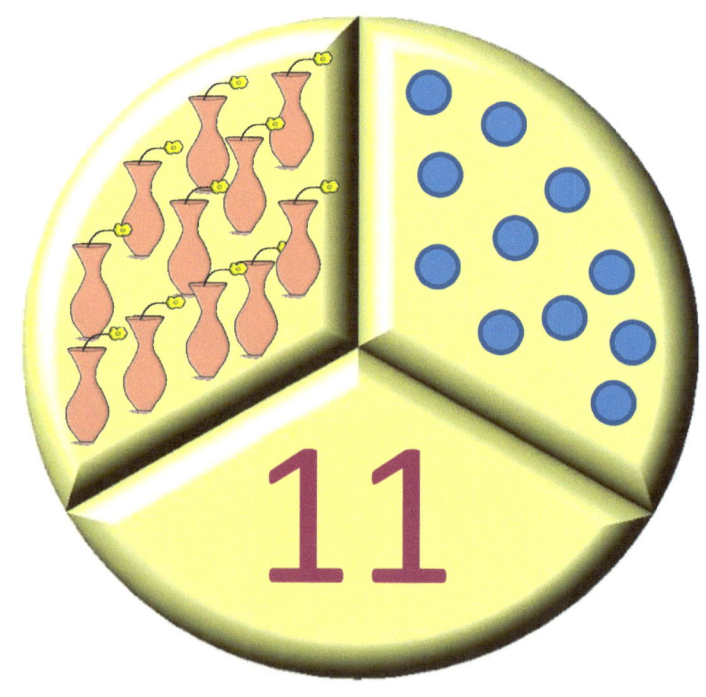

Fill in the missing number like in the big button below.

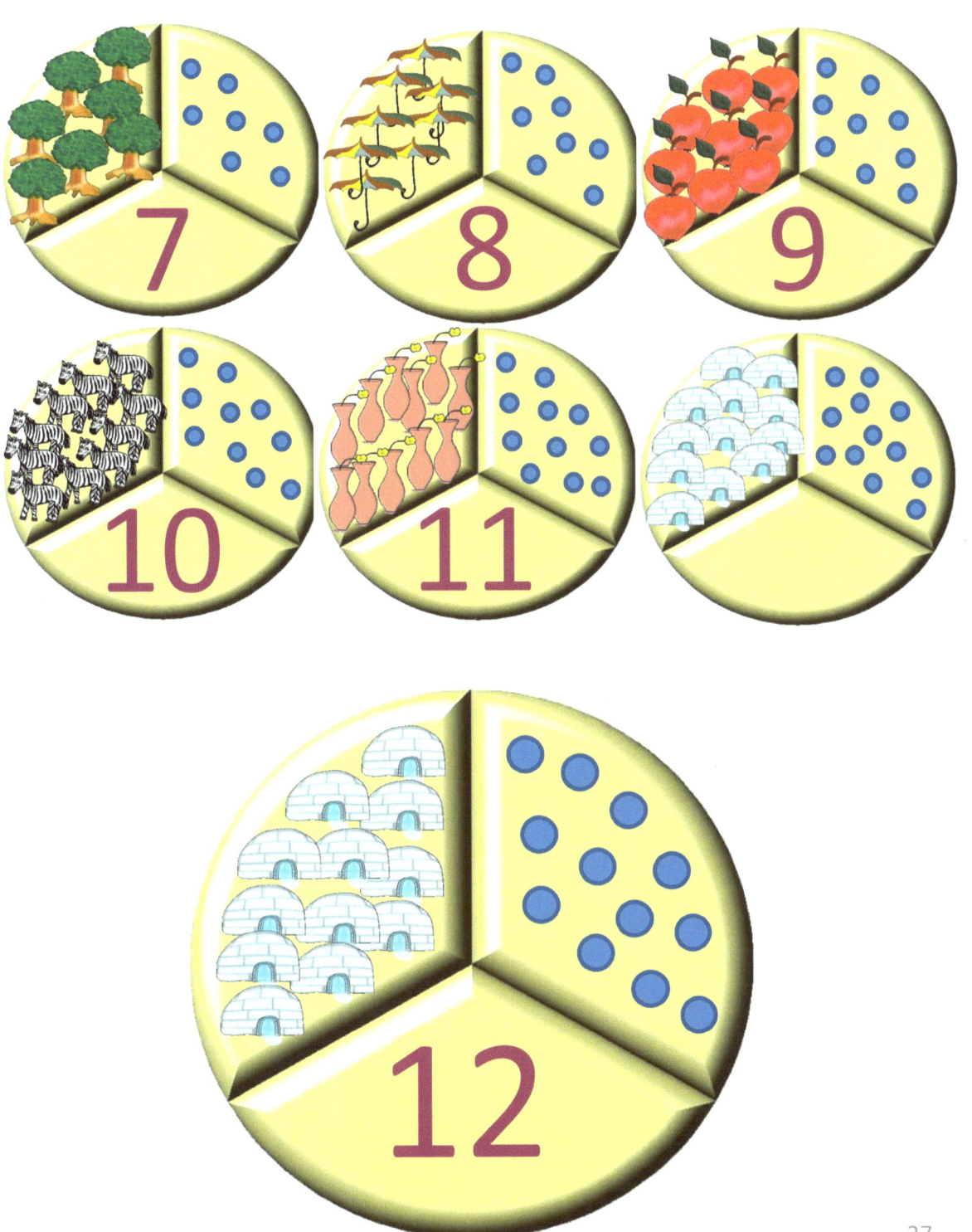

Fill in the missing numbers.

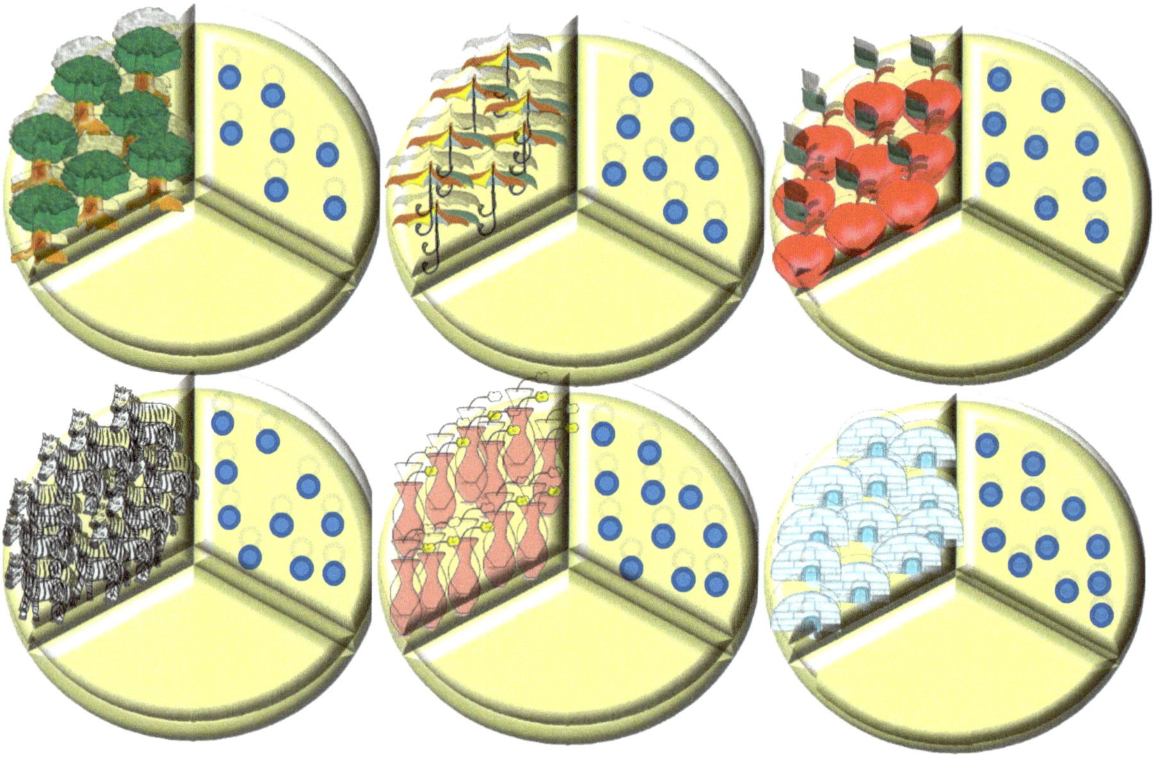

Fill in the missing number.

Fill in the missing numbers:

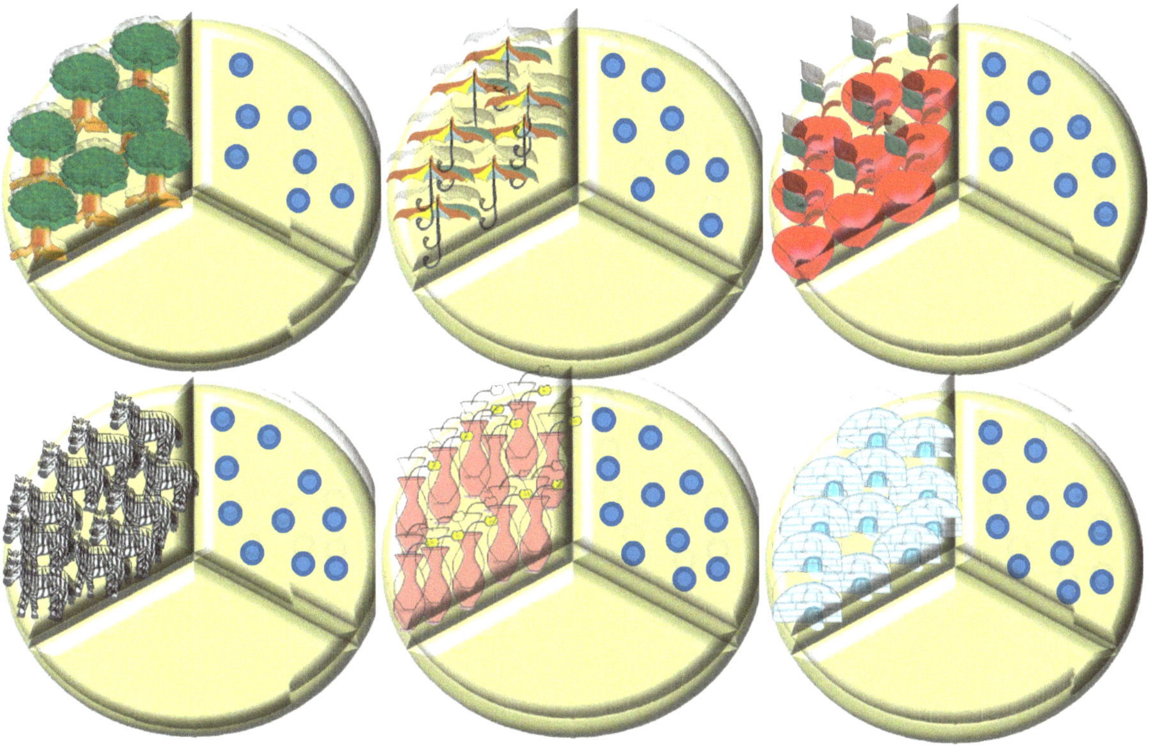

Fill in the missing number.

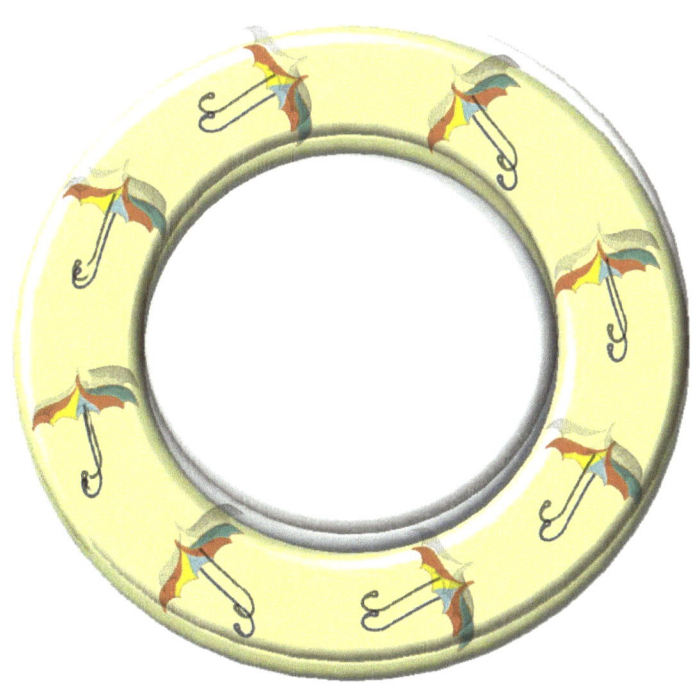

Fill in the missing numbers.

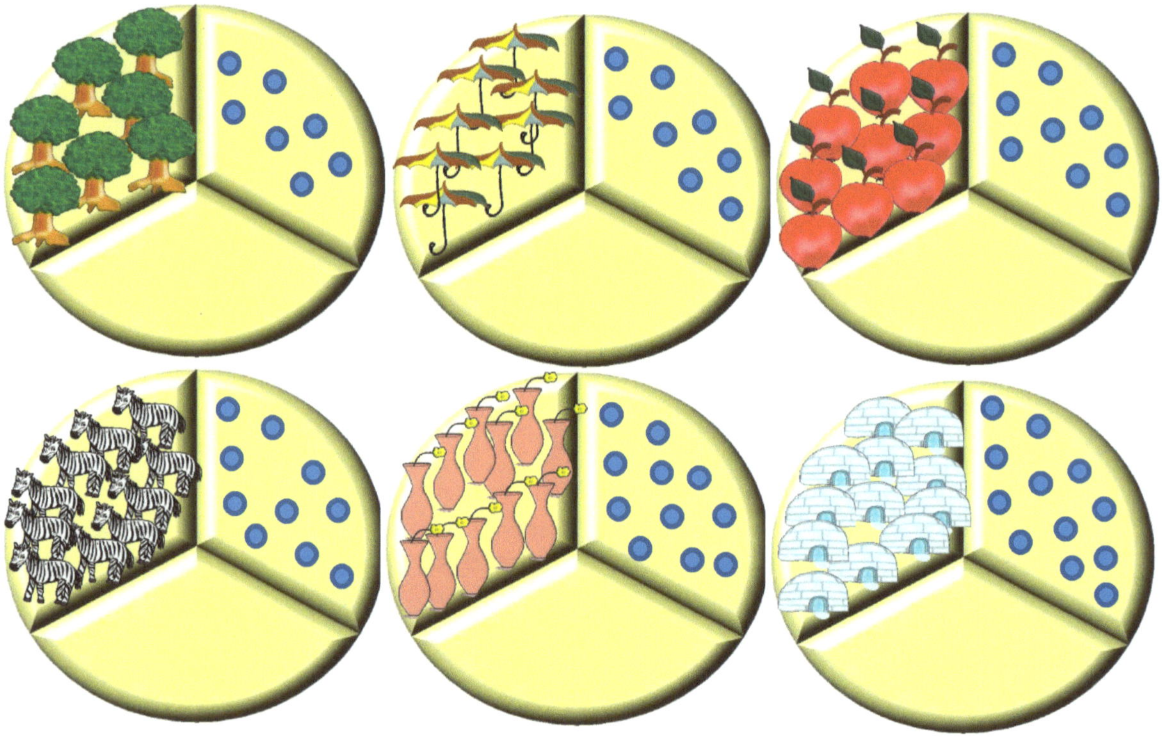

Fill in the missing number.

Fill in the missing numbers.

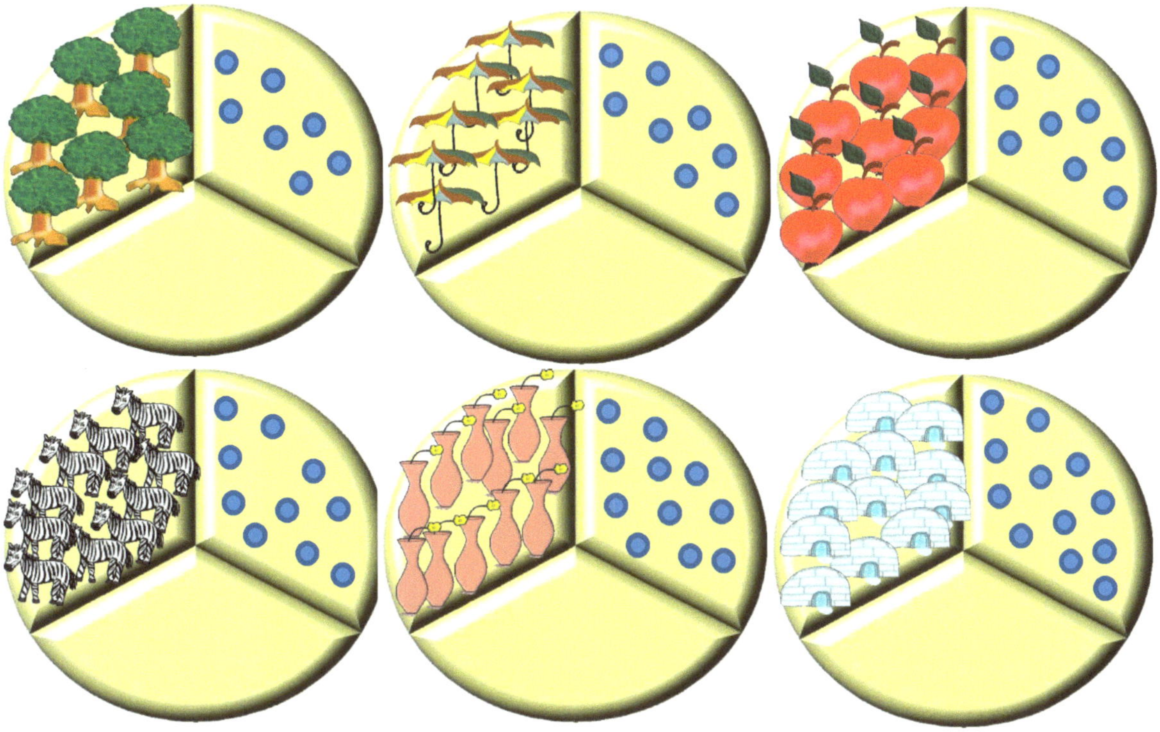

Fill in the missing number.

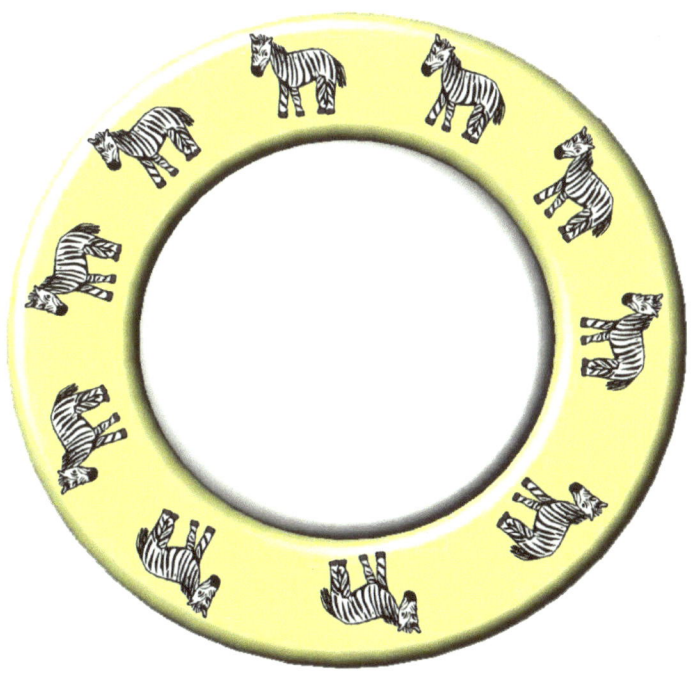

Fill in the missing numbers.

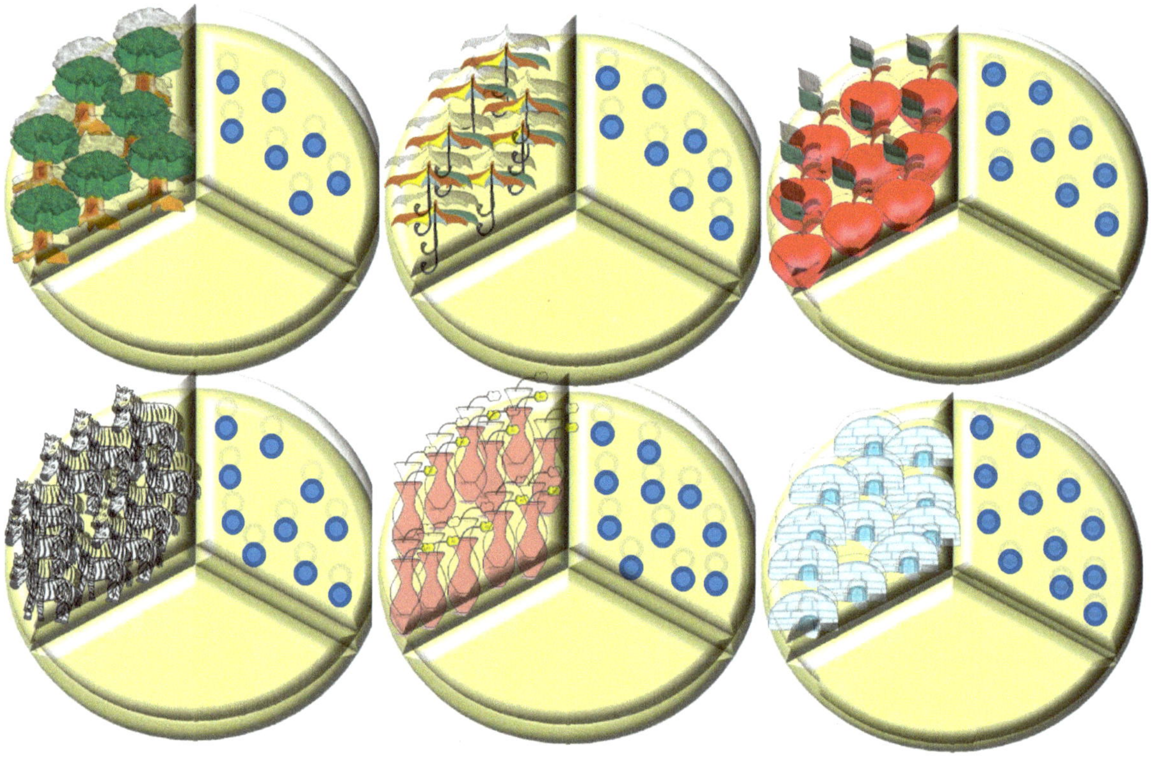

Fill in the missing number.

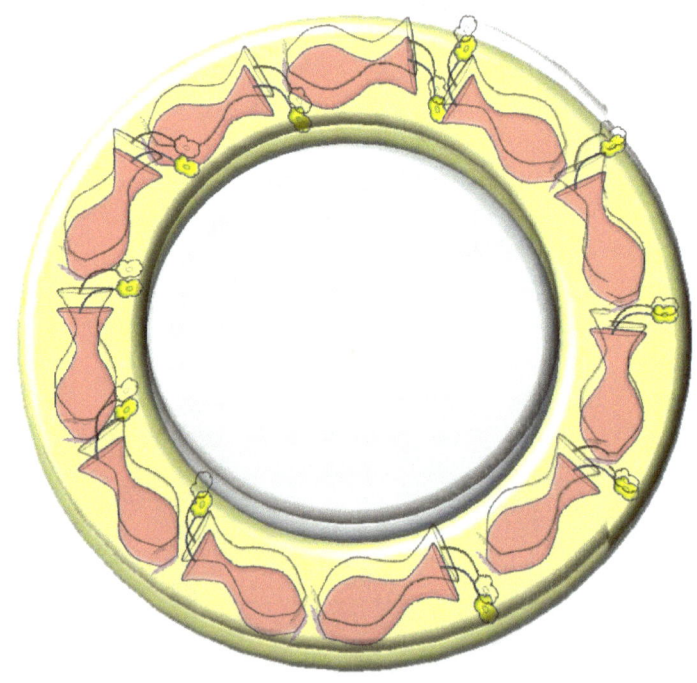

Fill in the missing numbers:

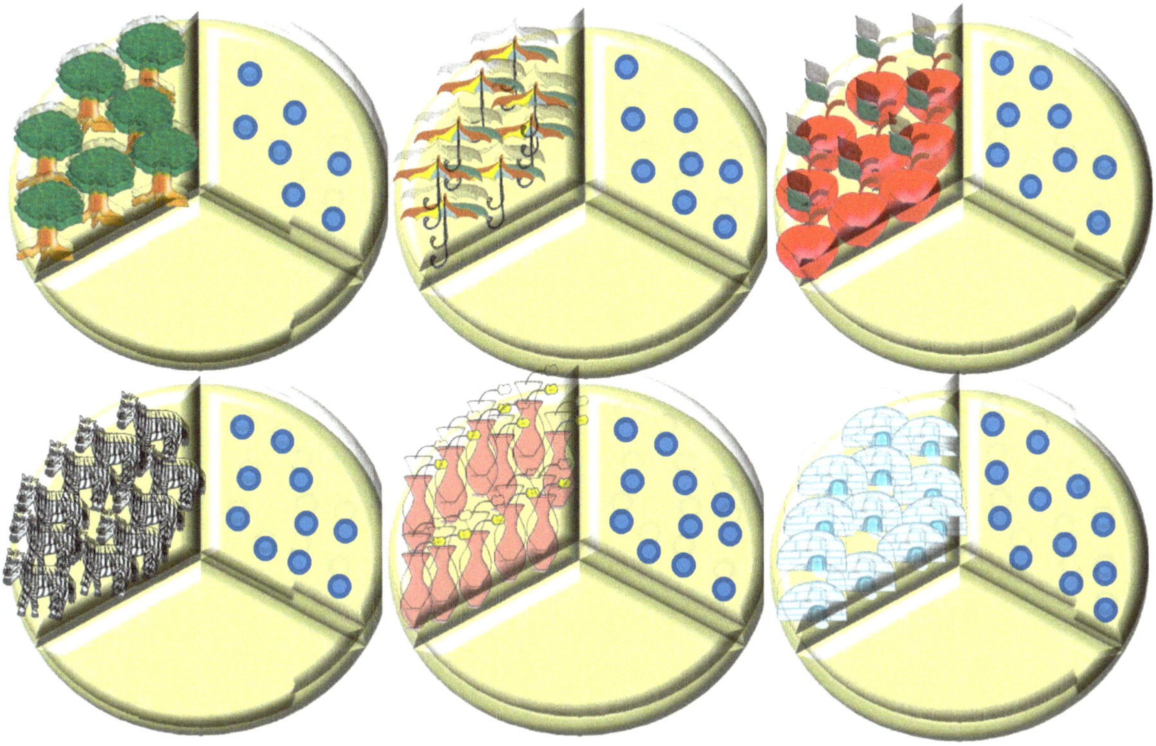

Fill in the missing number.

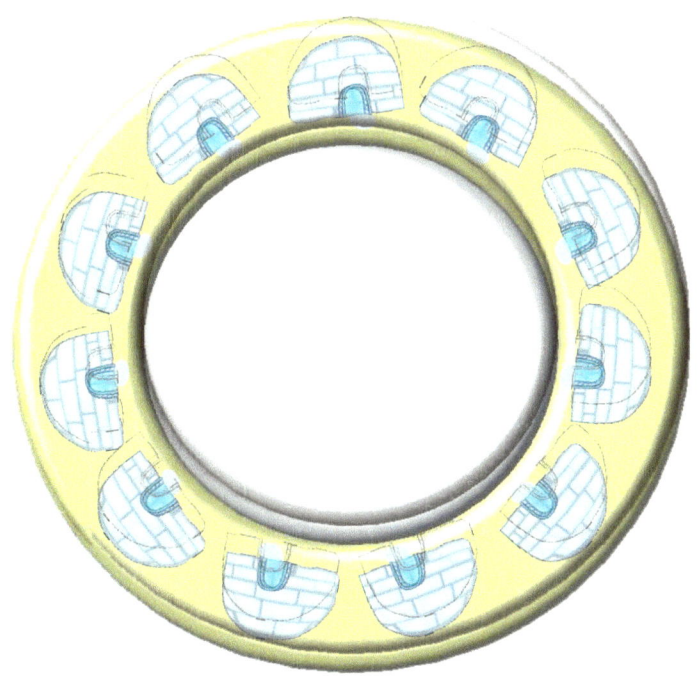

Fill in the missing numbers.

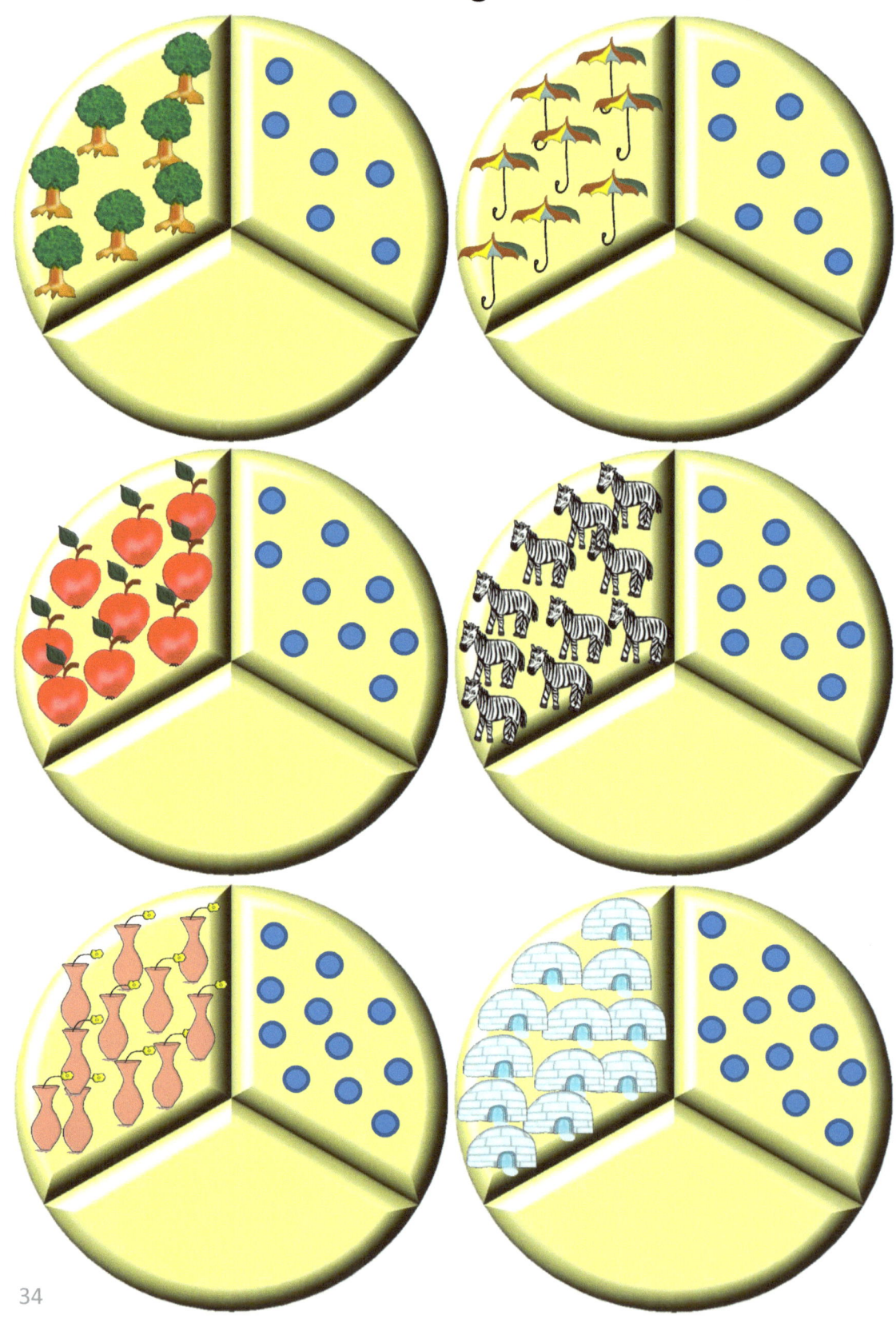

Fill in the missing dots and numbers.

Fill in the missing numbers.

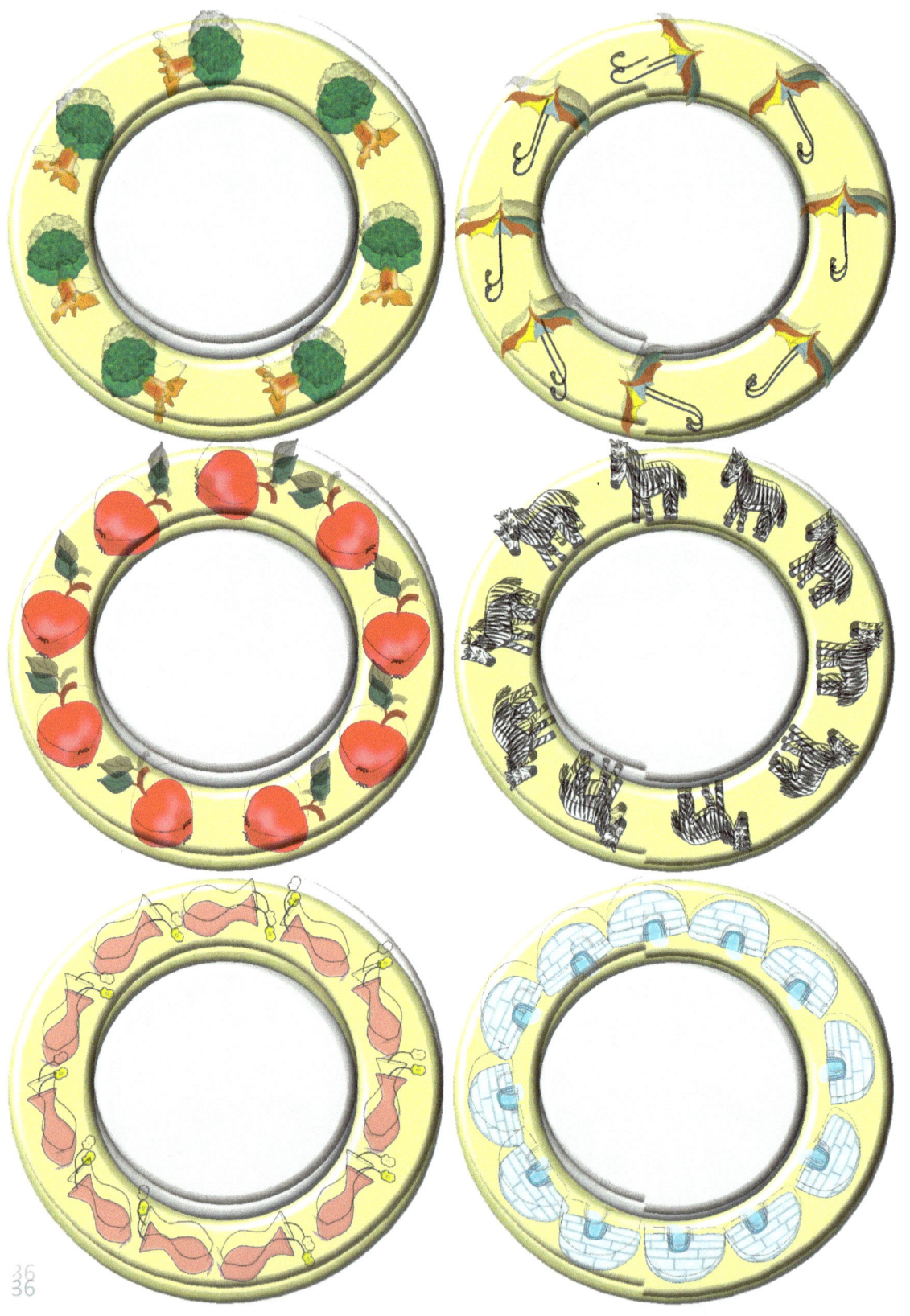

Fill in the missing numbers:

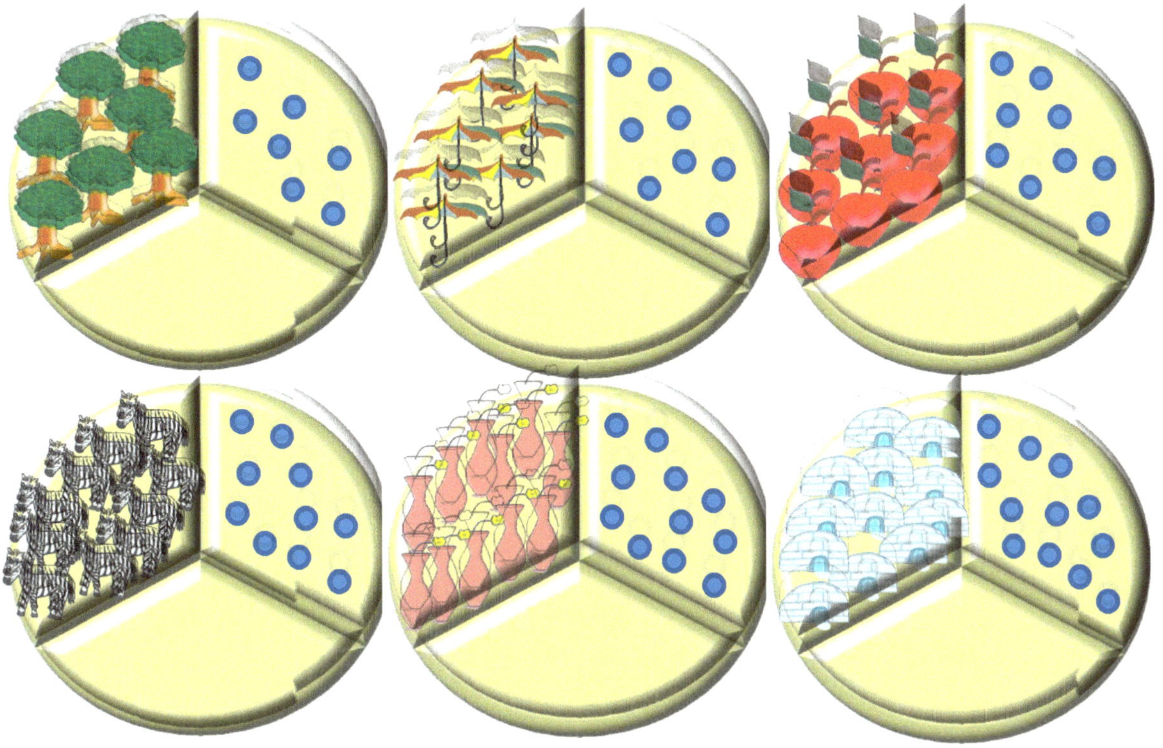

Fill in the missing numbers:

Fill in the missing dots and numbers.

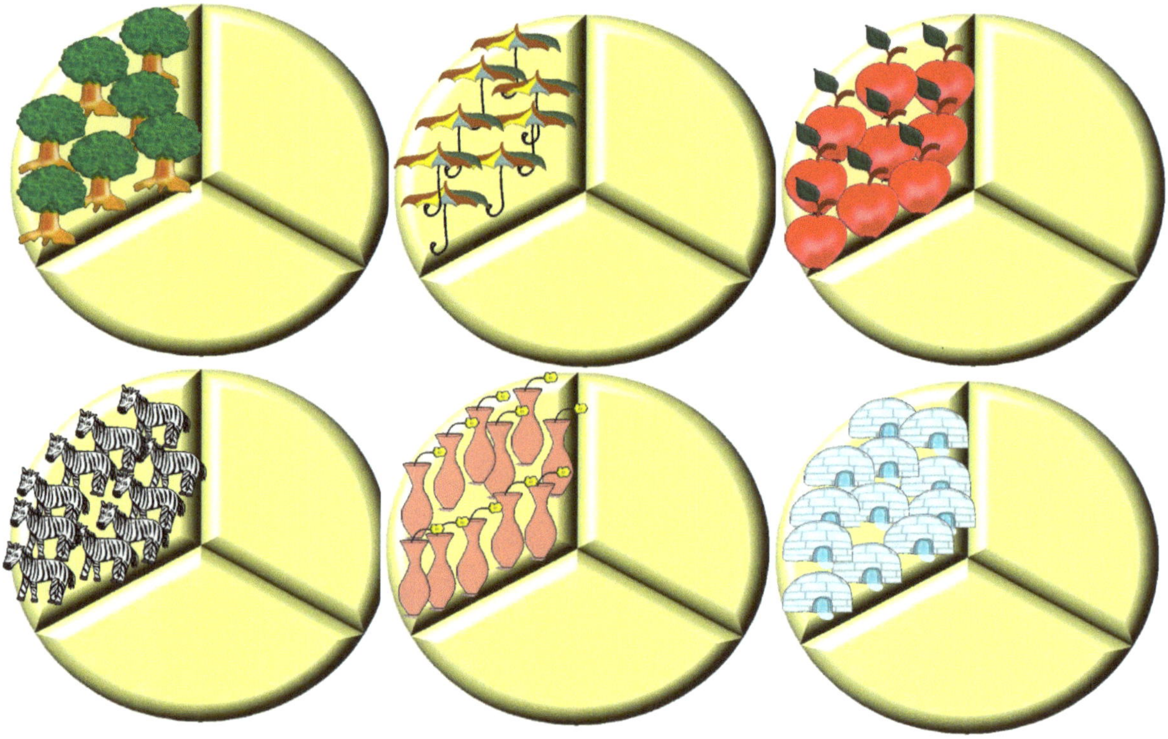

Fill in the missing numbers.

1-12 numbers chart.

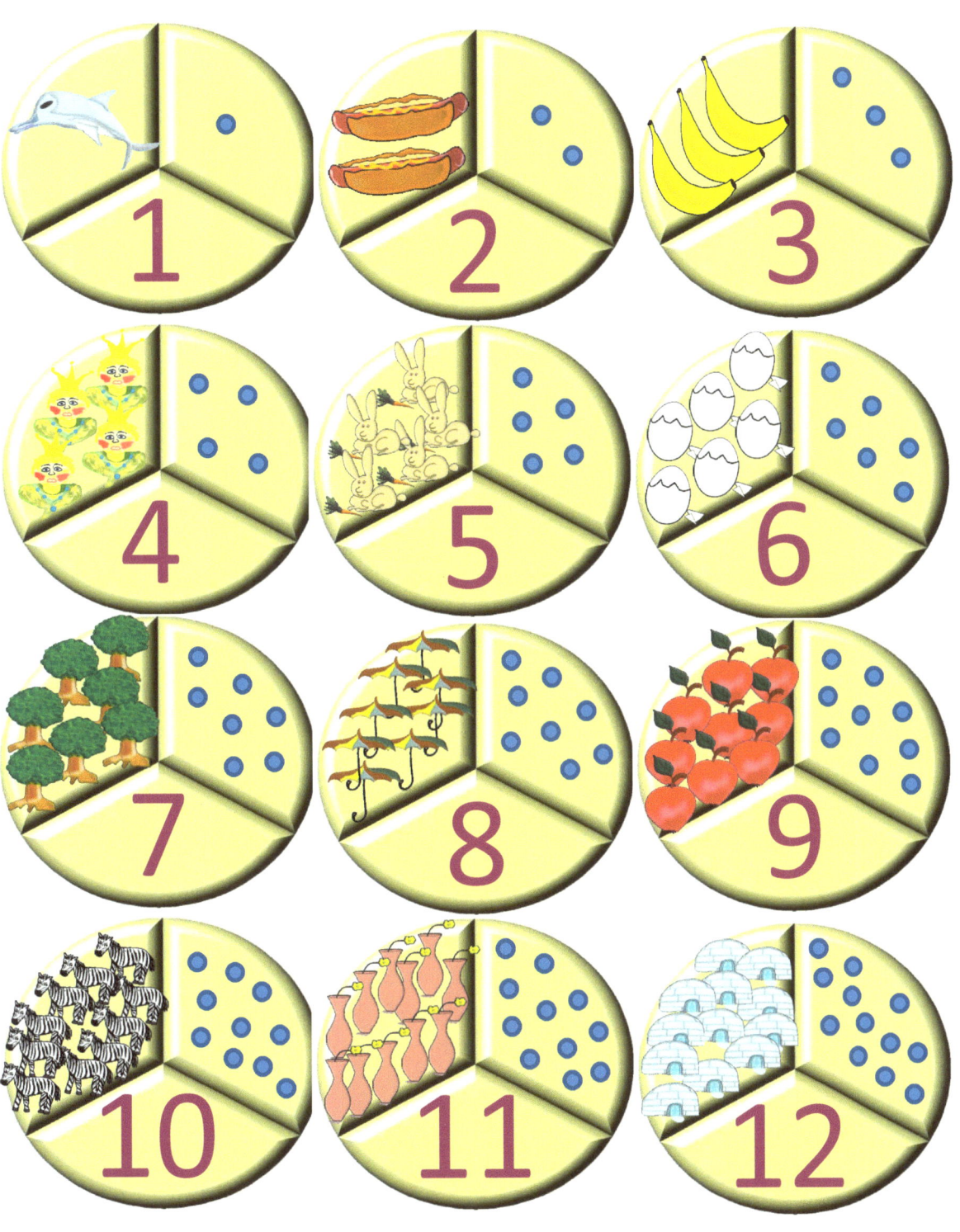

Fill in the missing numbers.

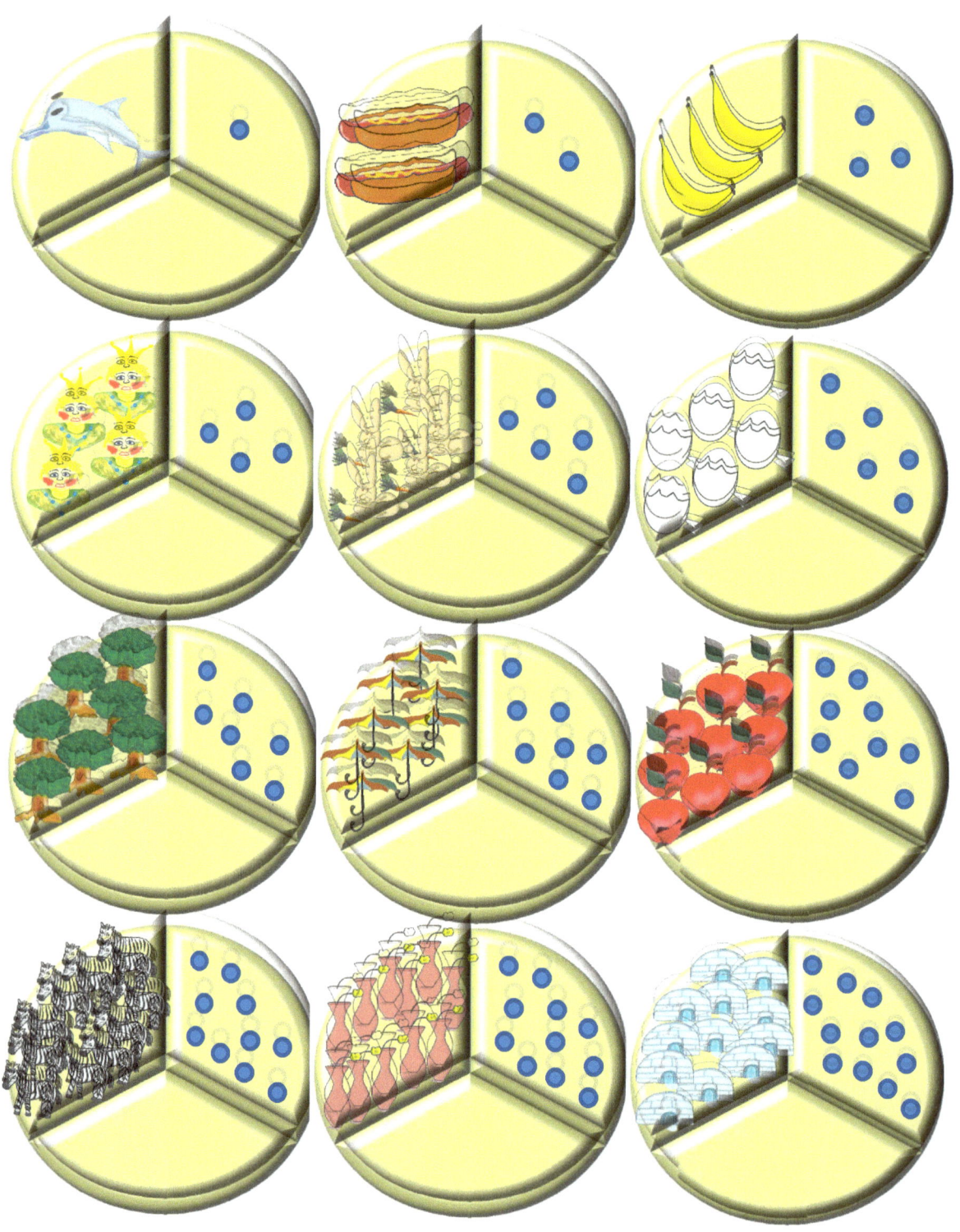

Fill in the missing dots and numbers:

Fill in the missing numbers.

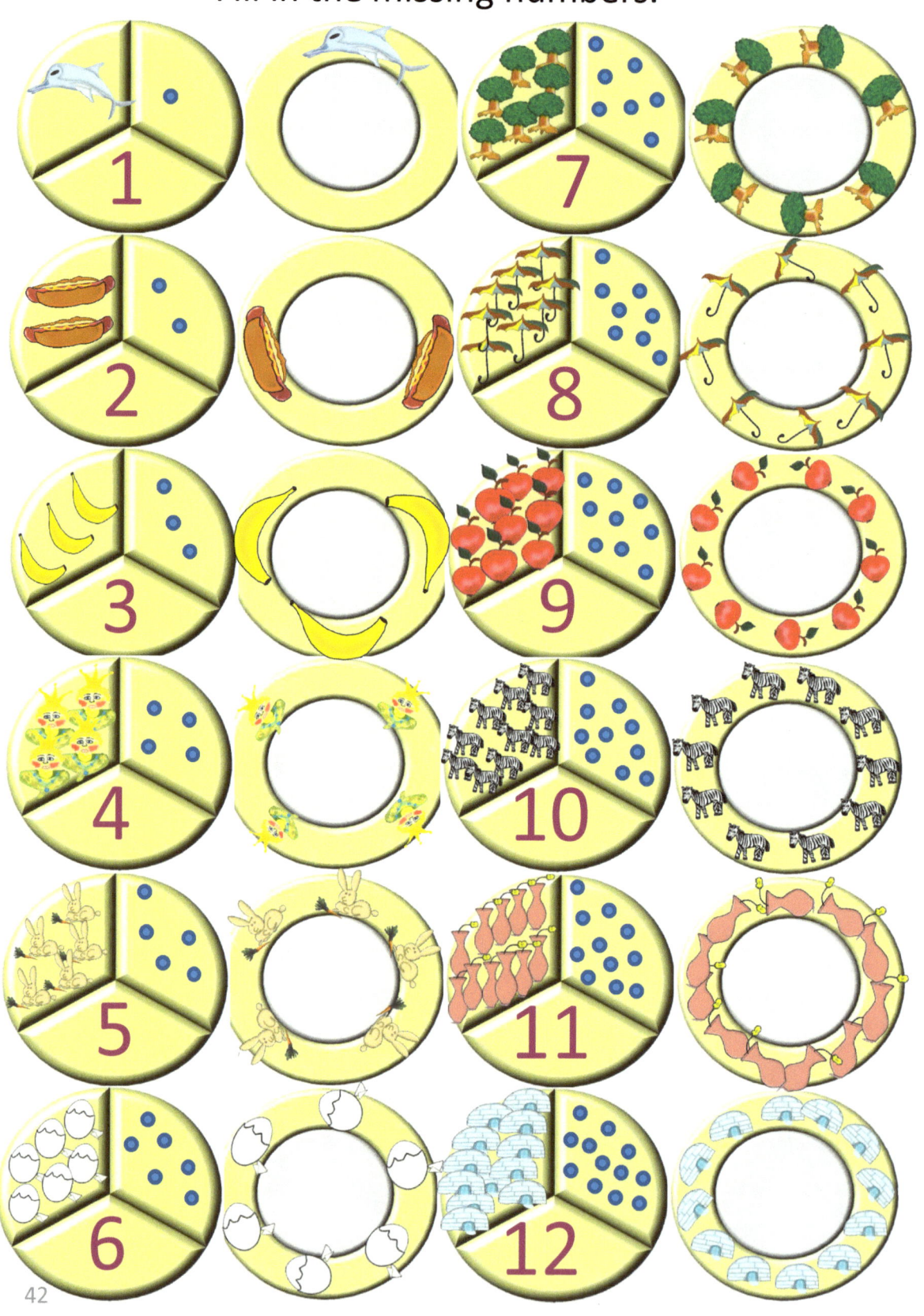

Fill in the missing numbers.

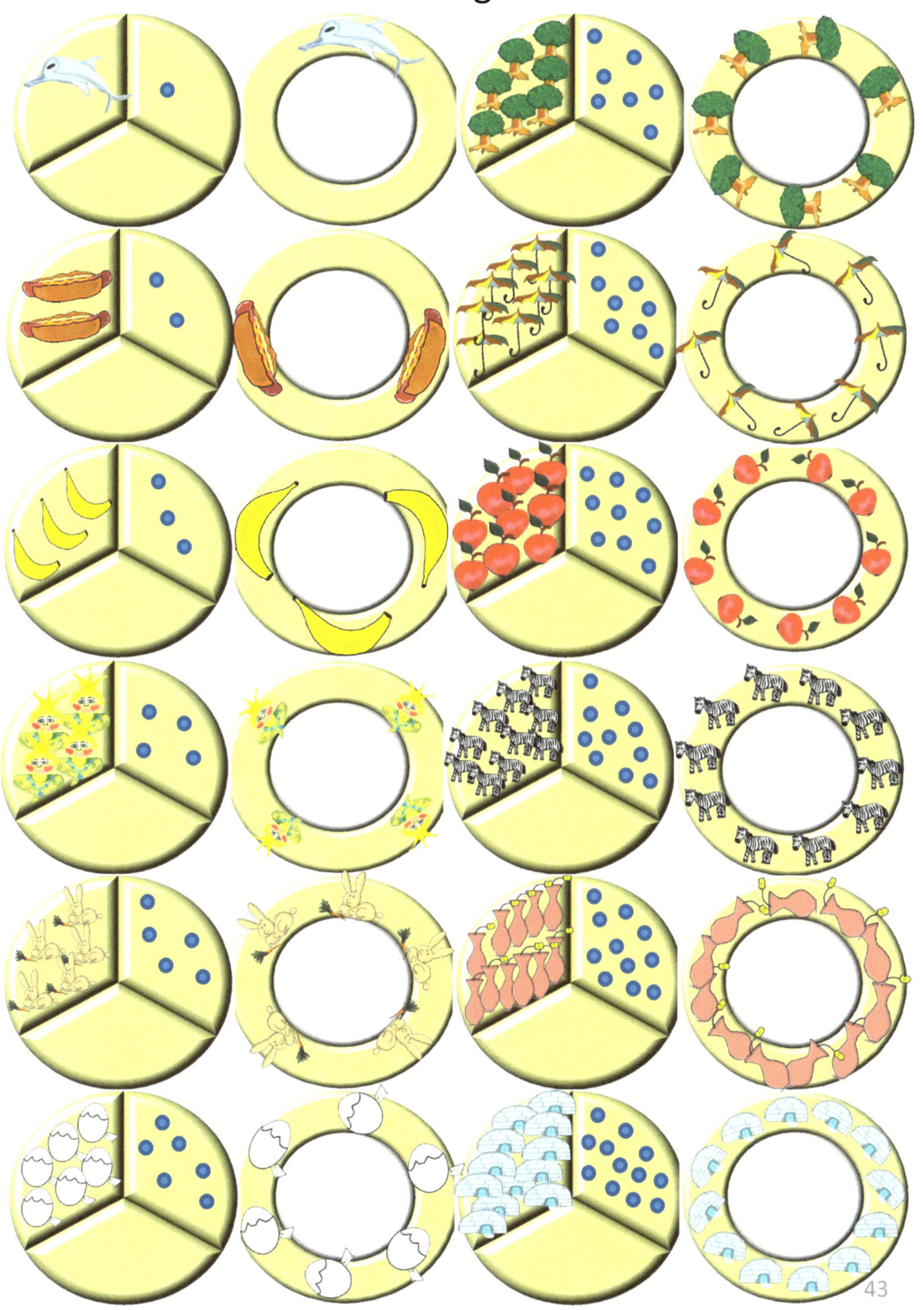

Fill in the missing dots and numbers.

Fill in the missing numbers 1 – 12:

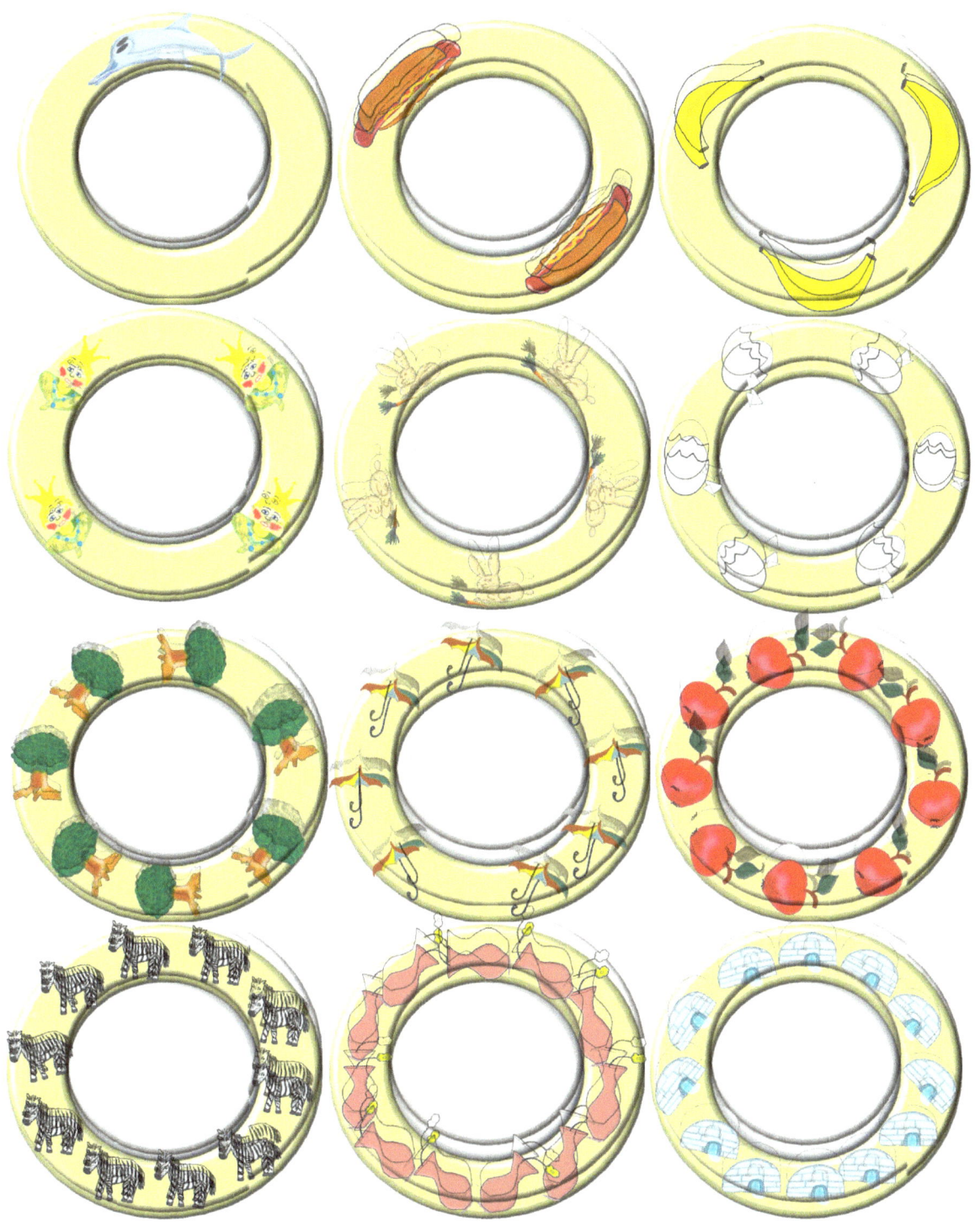

Addition exercises.

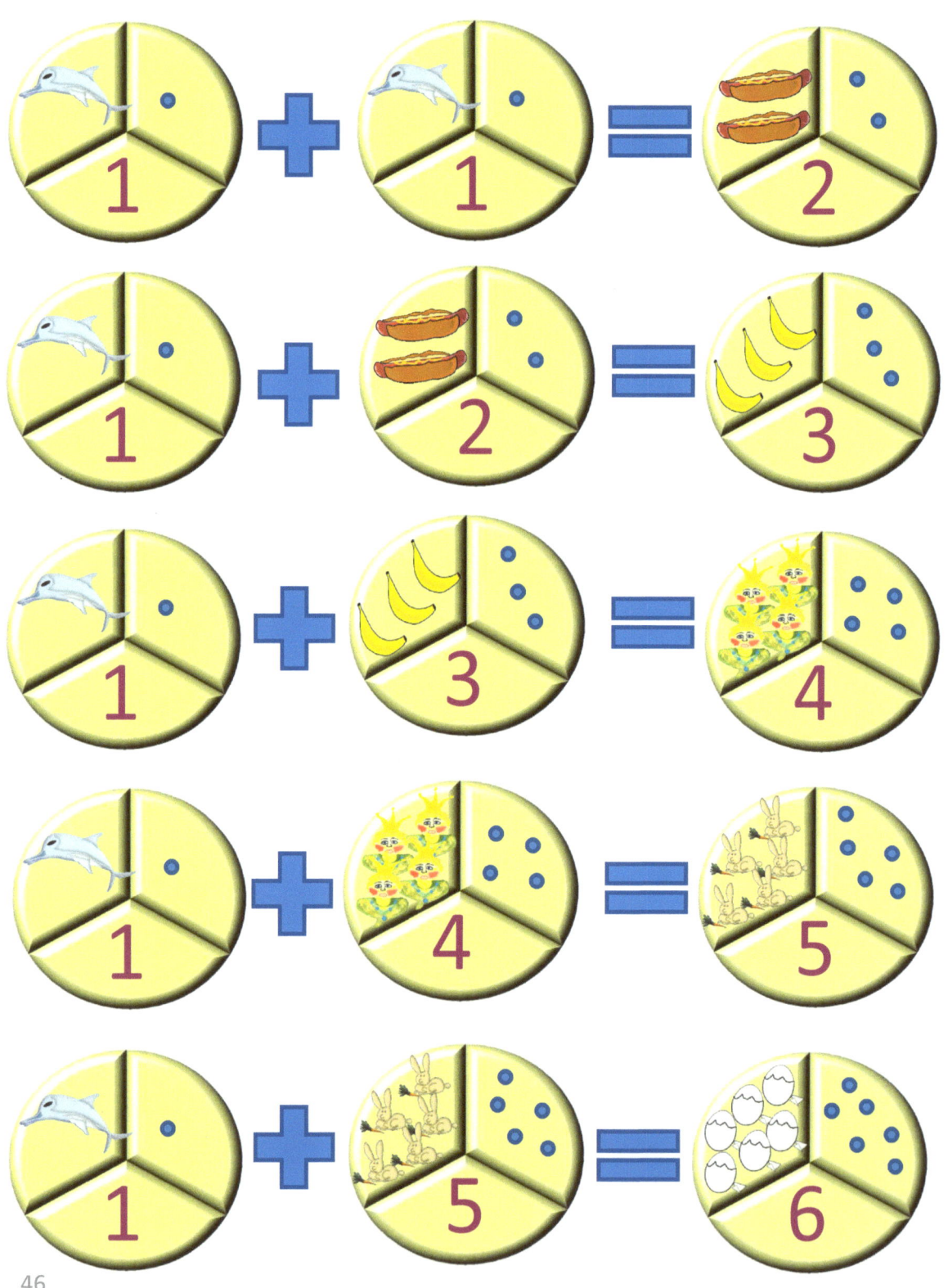

Fill in the missing numbers.

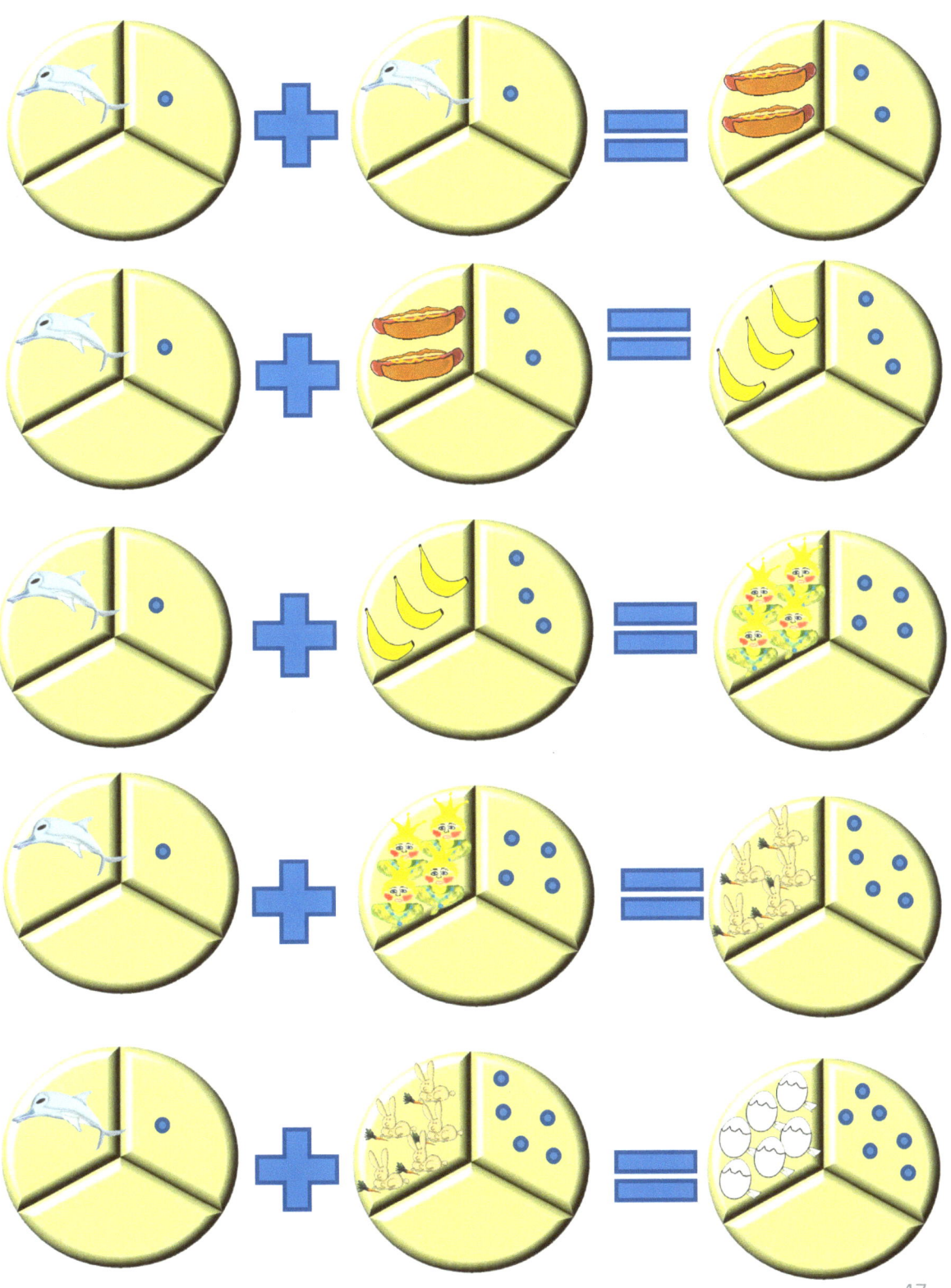

Fill in the missing numbers.

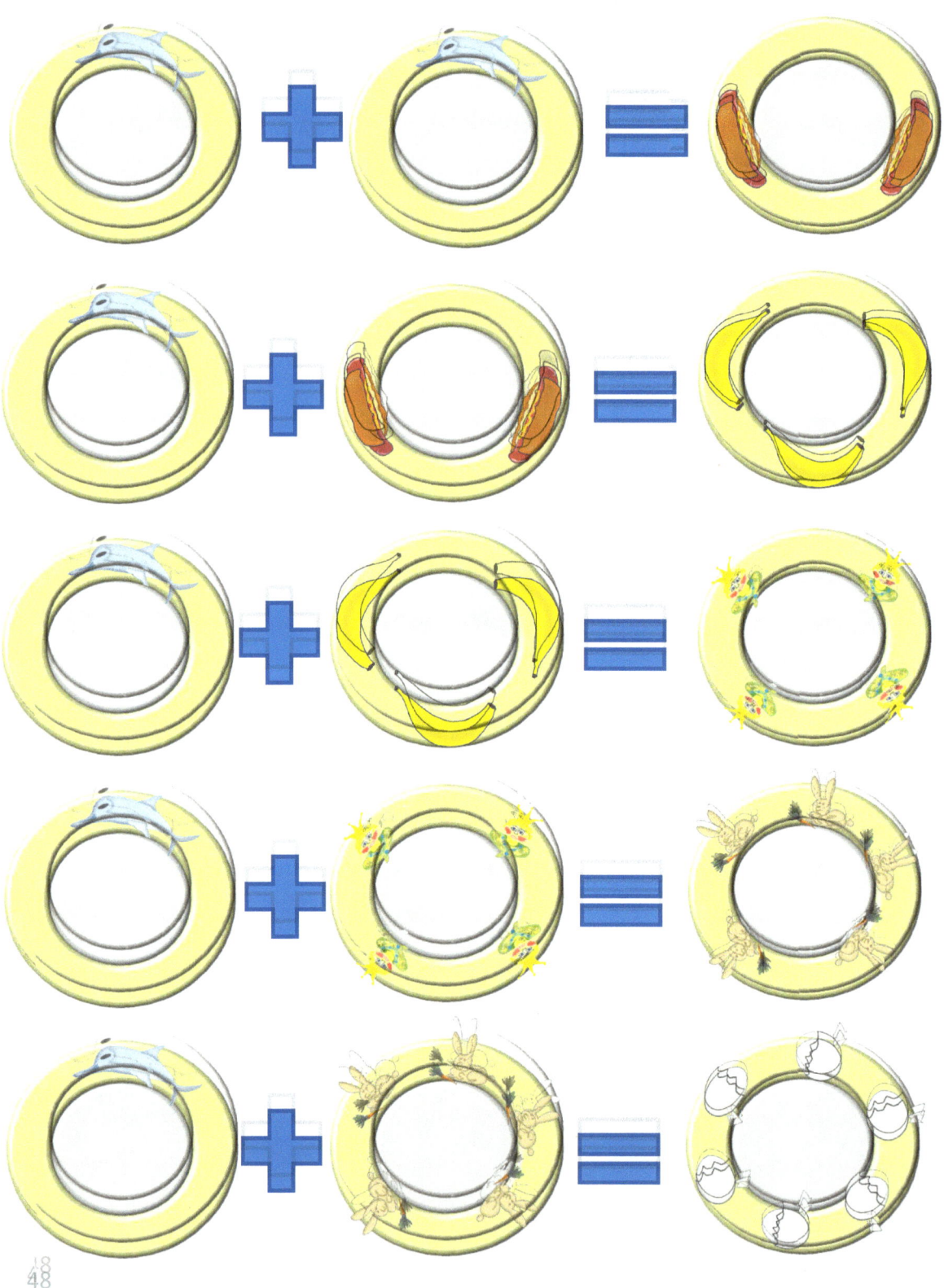

Addition exercises:

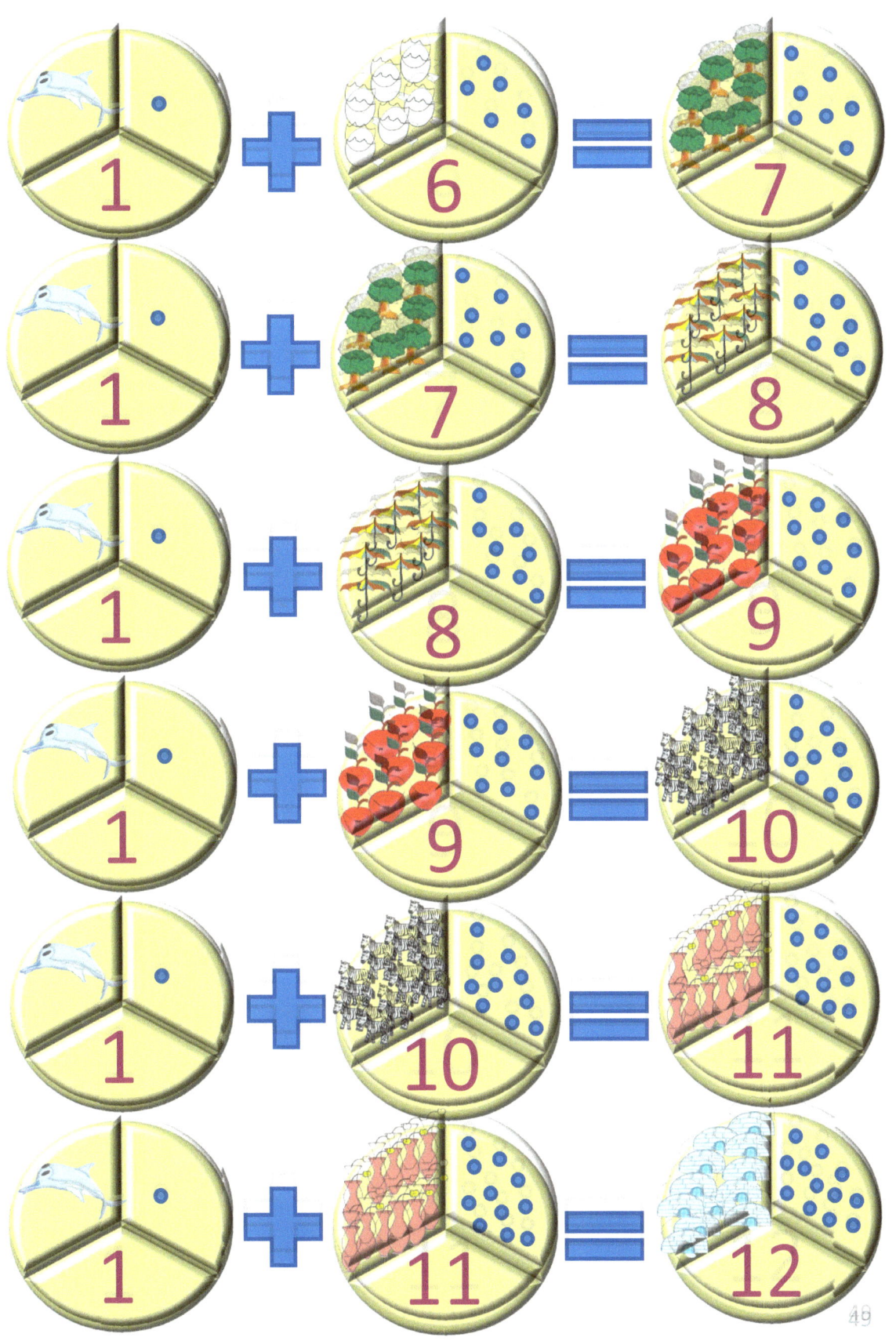

Fill in the missing numbers.

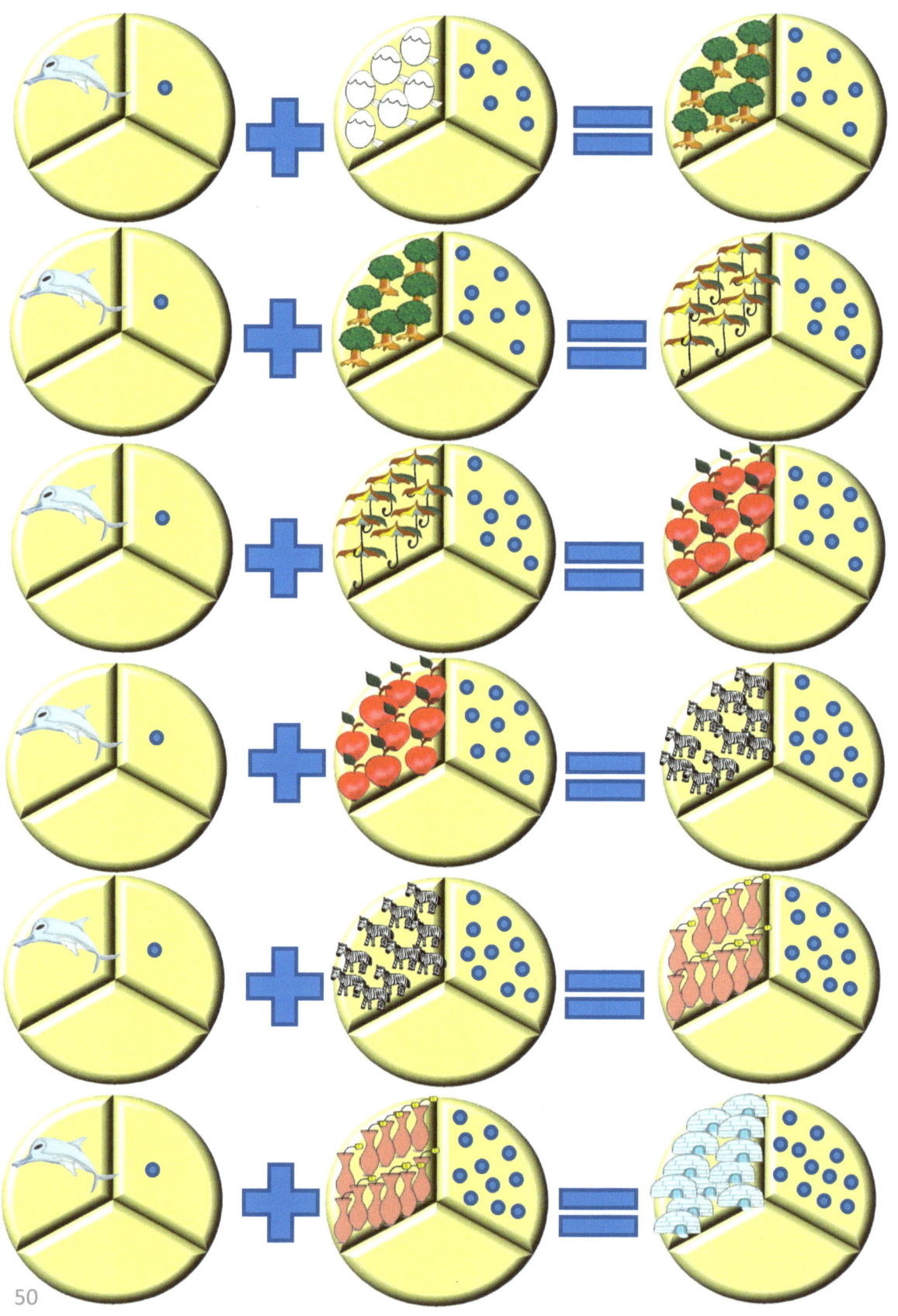

Fill in the missing numbers.

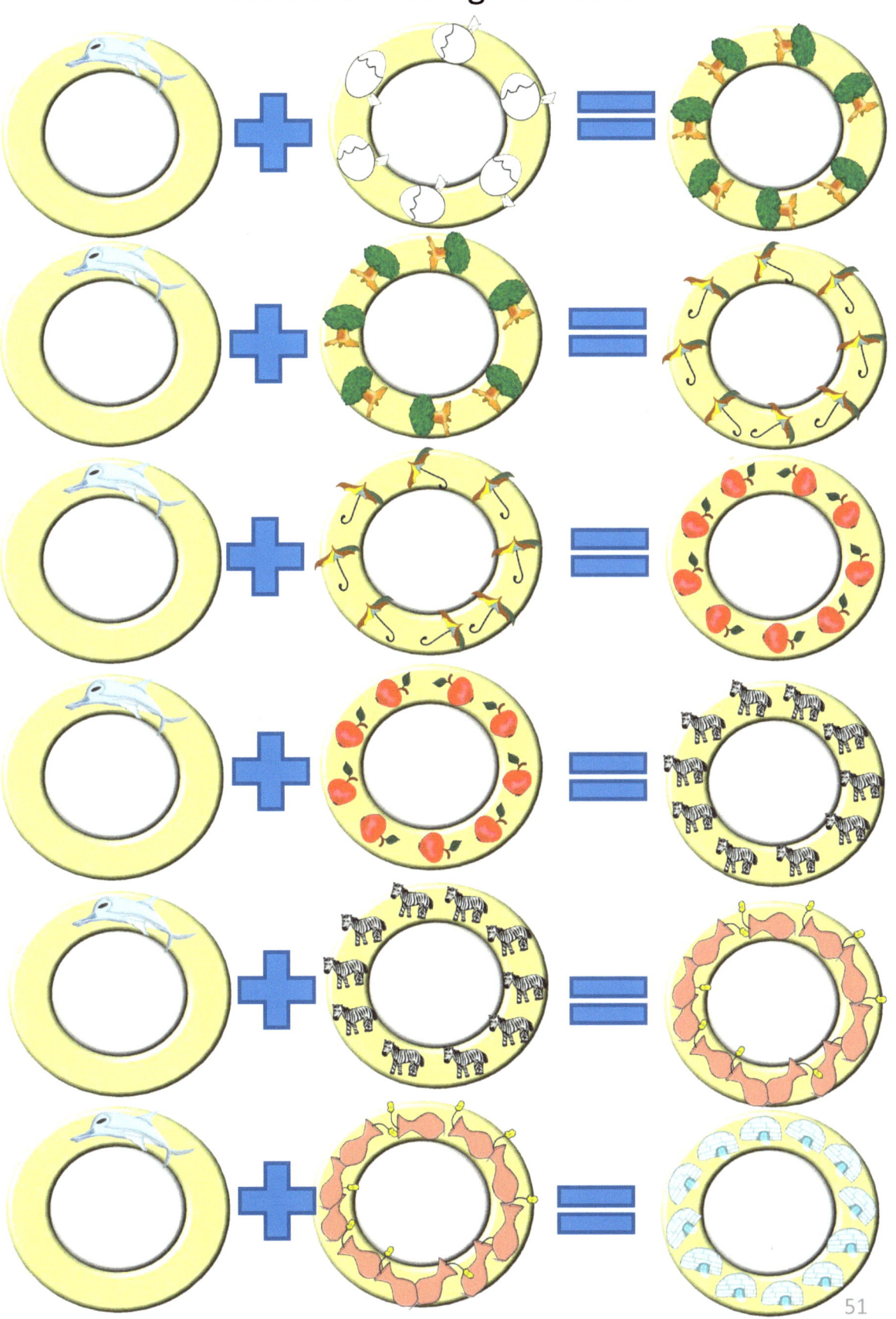

Addition exercises:

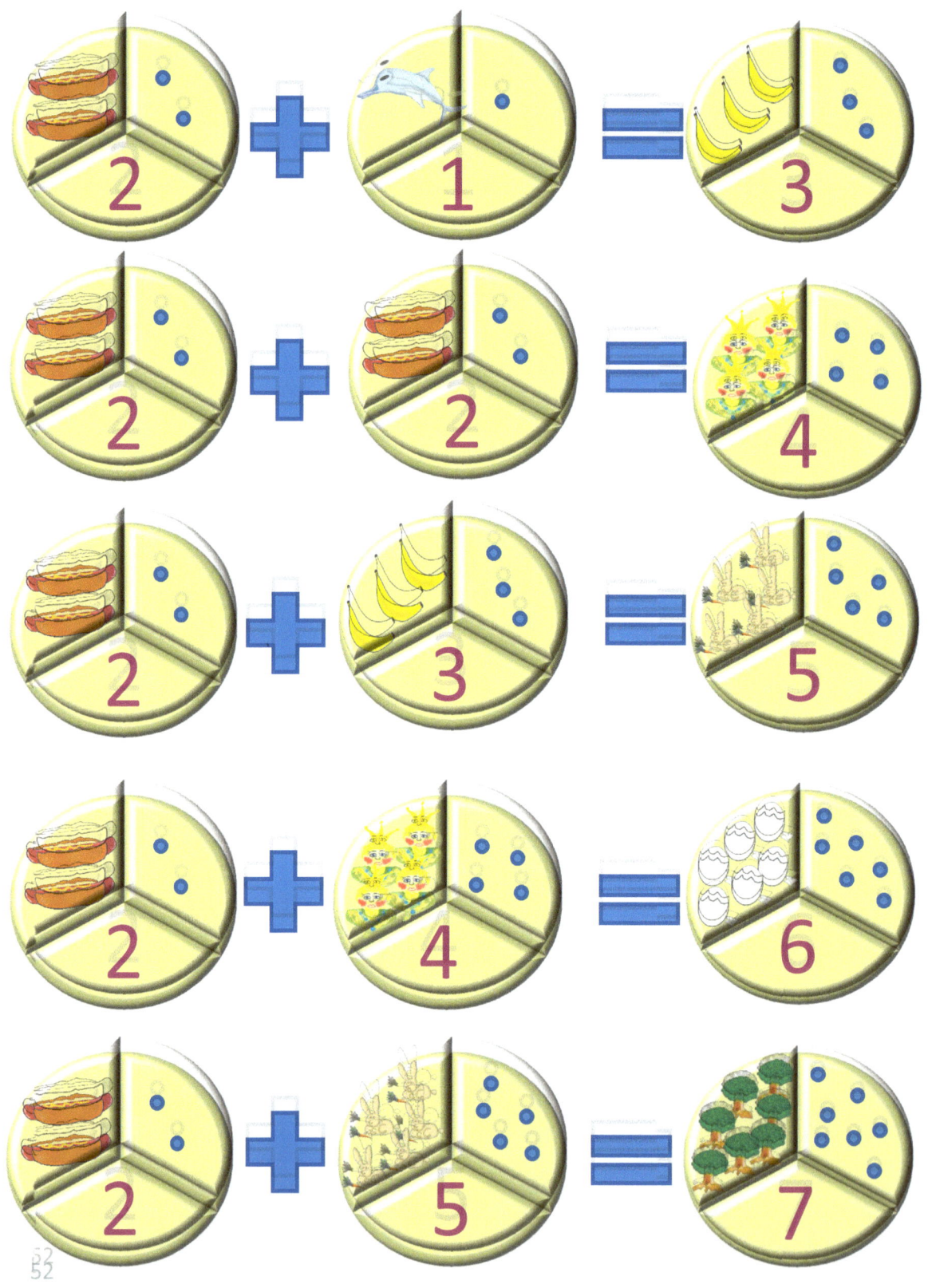

Fill in the missing numbers:

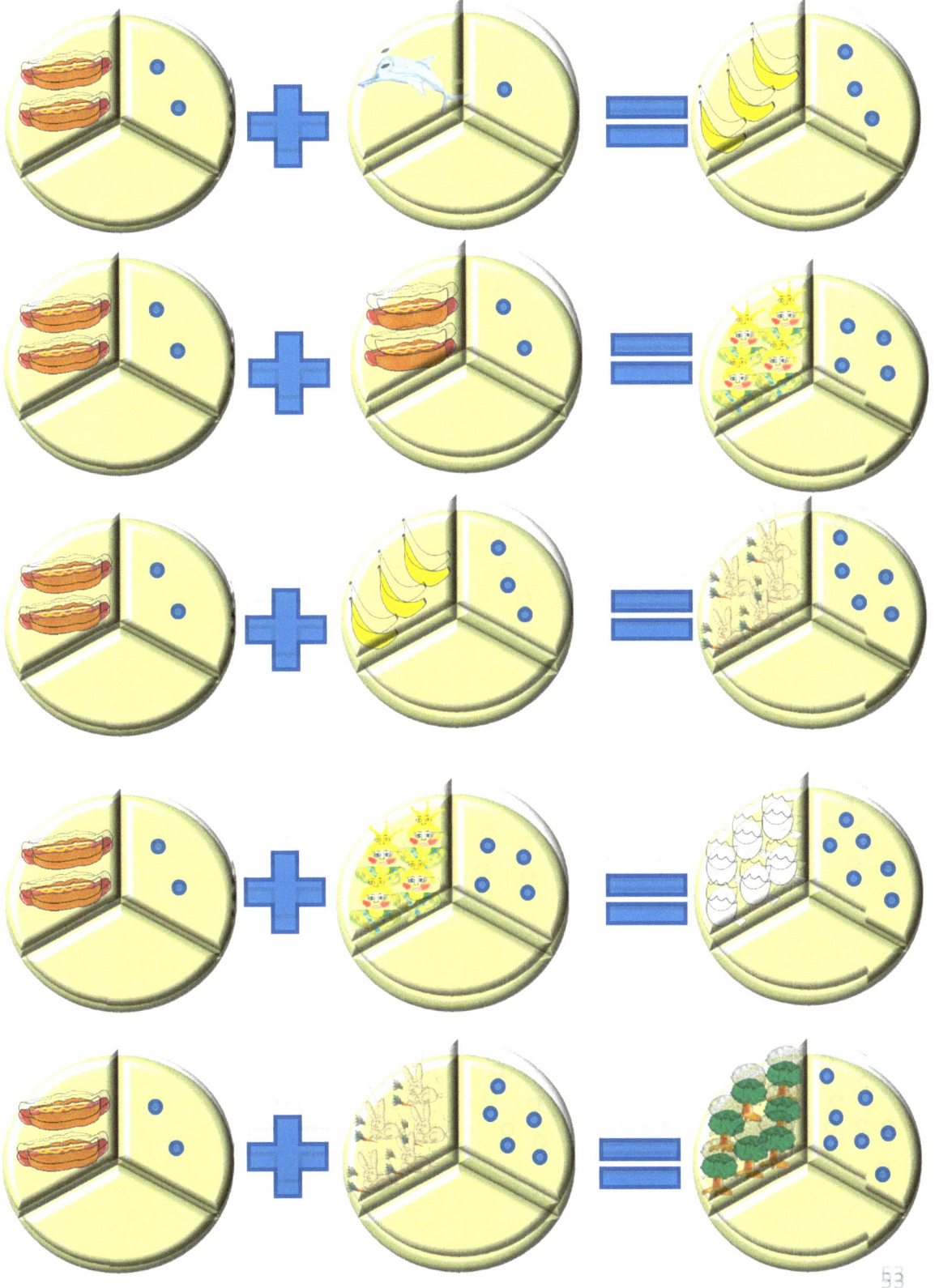

Fill in the missing numbers.

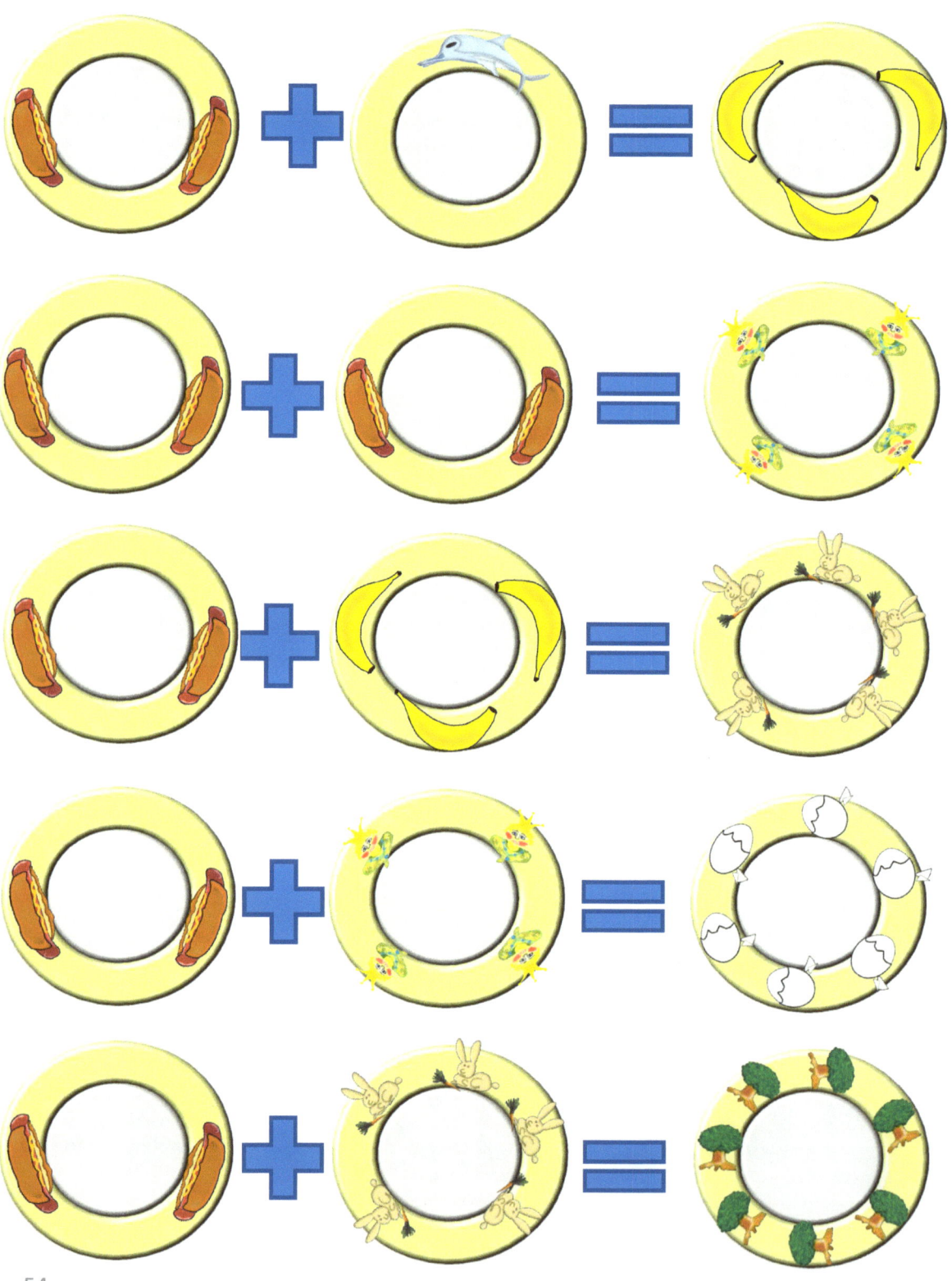

Addition exercises.

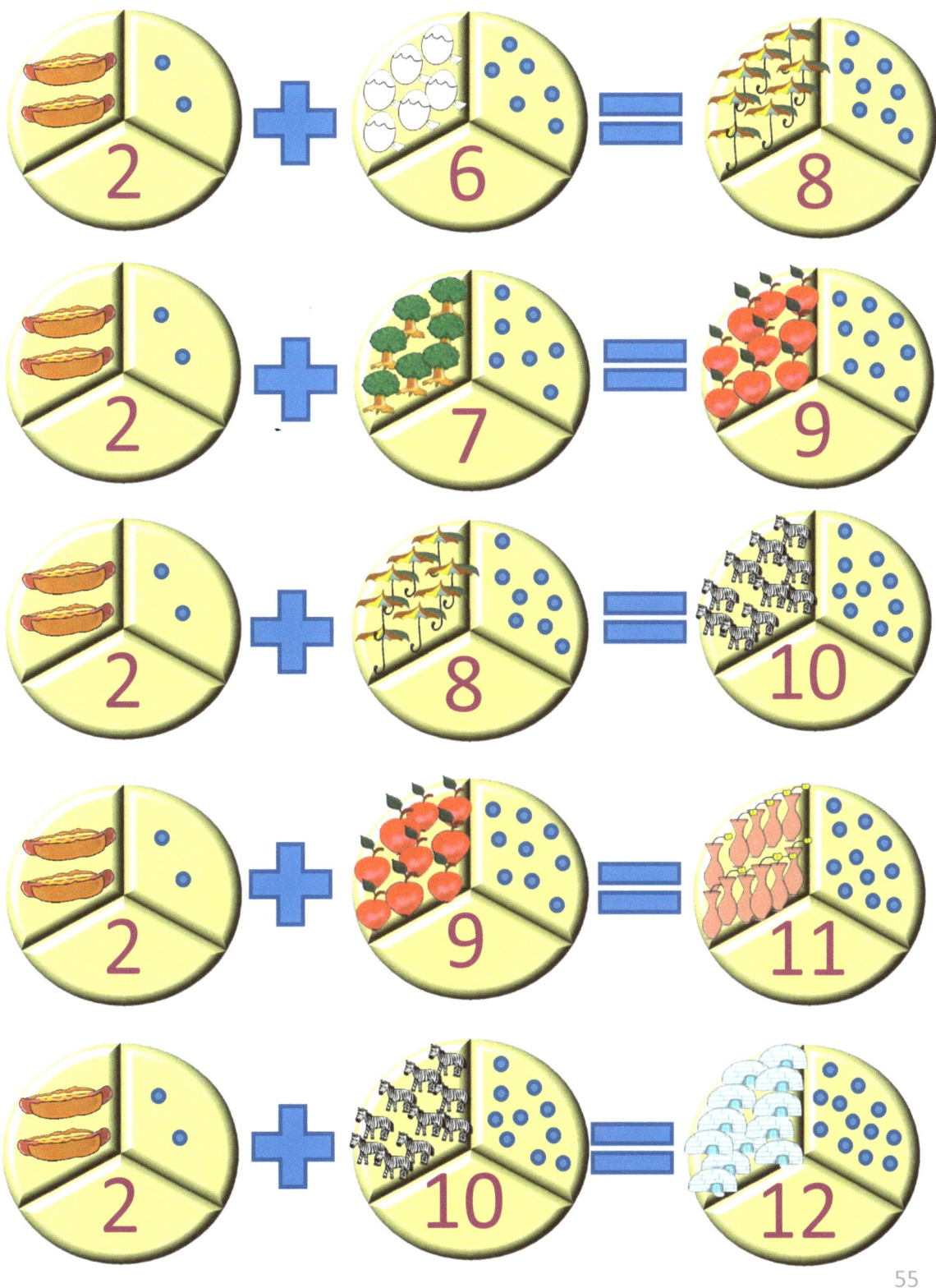

Fill in the missing numbers.

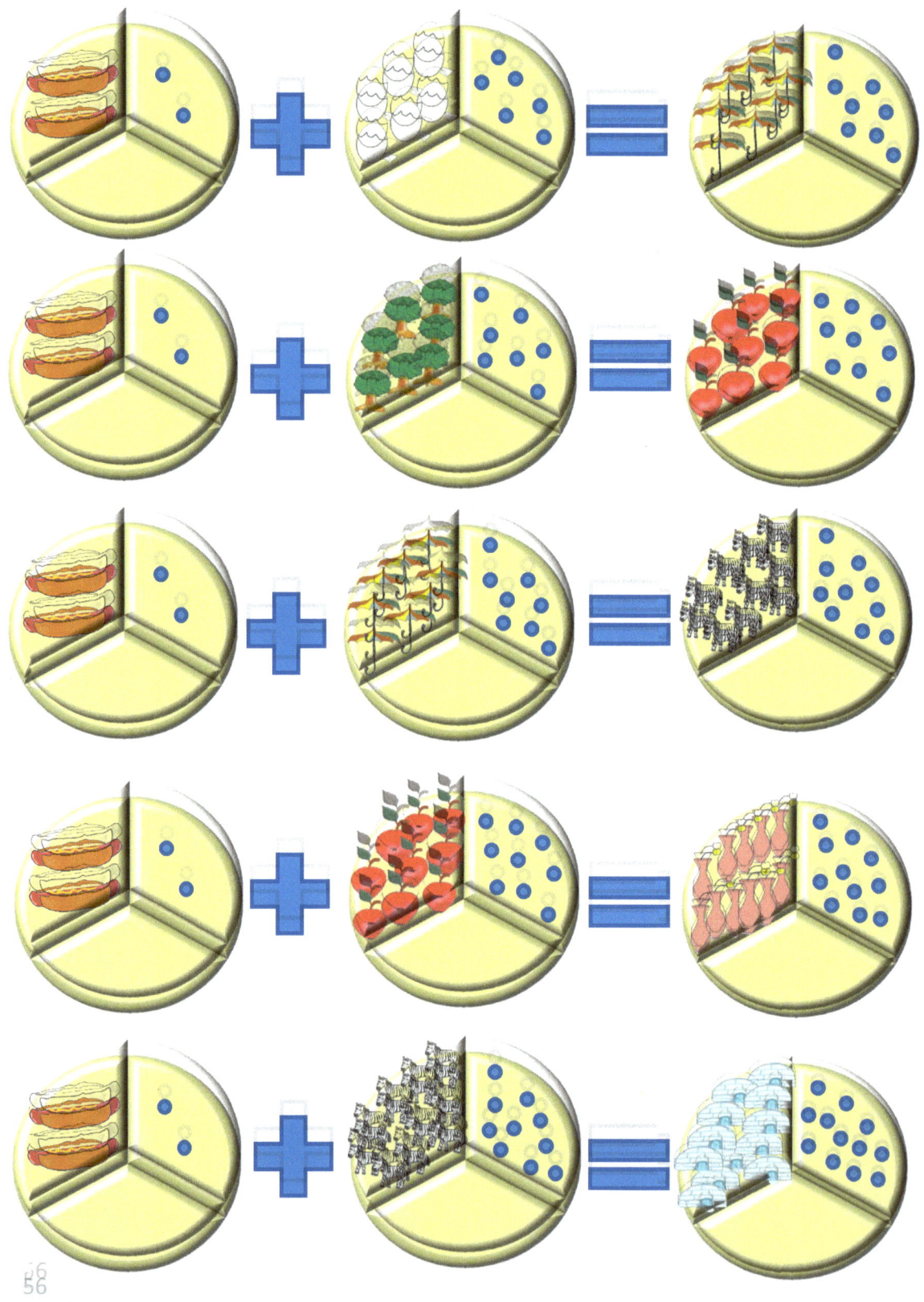

fill in the missing numbers:

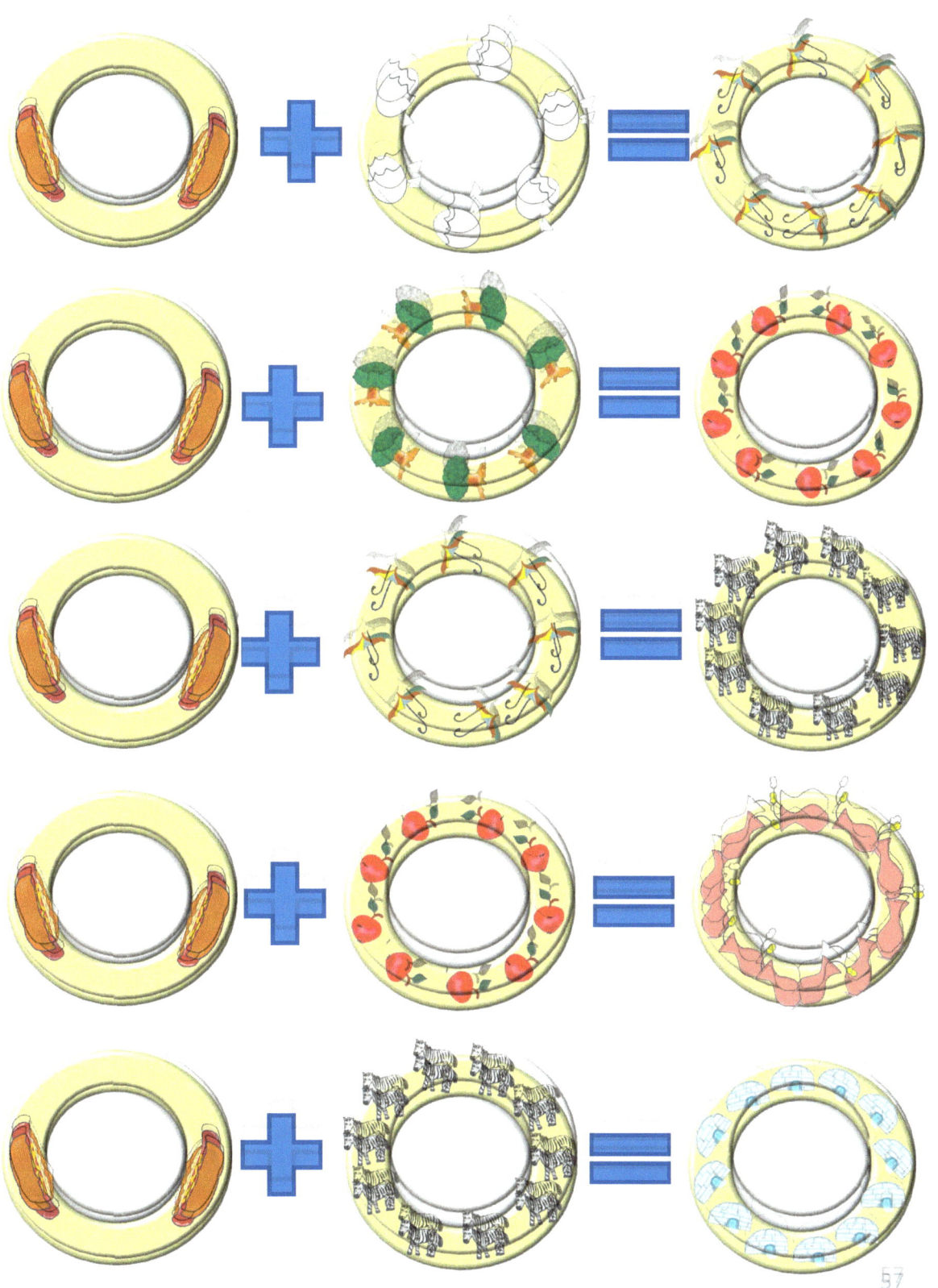

Fill in the missing numbers 1-12.

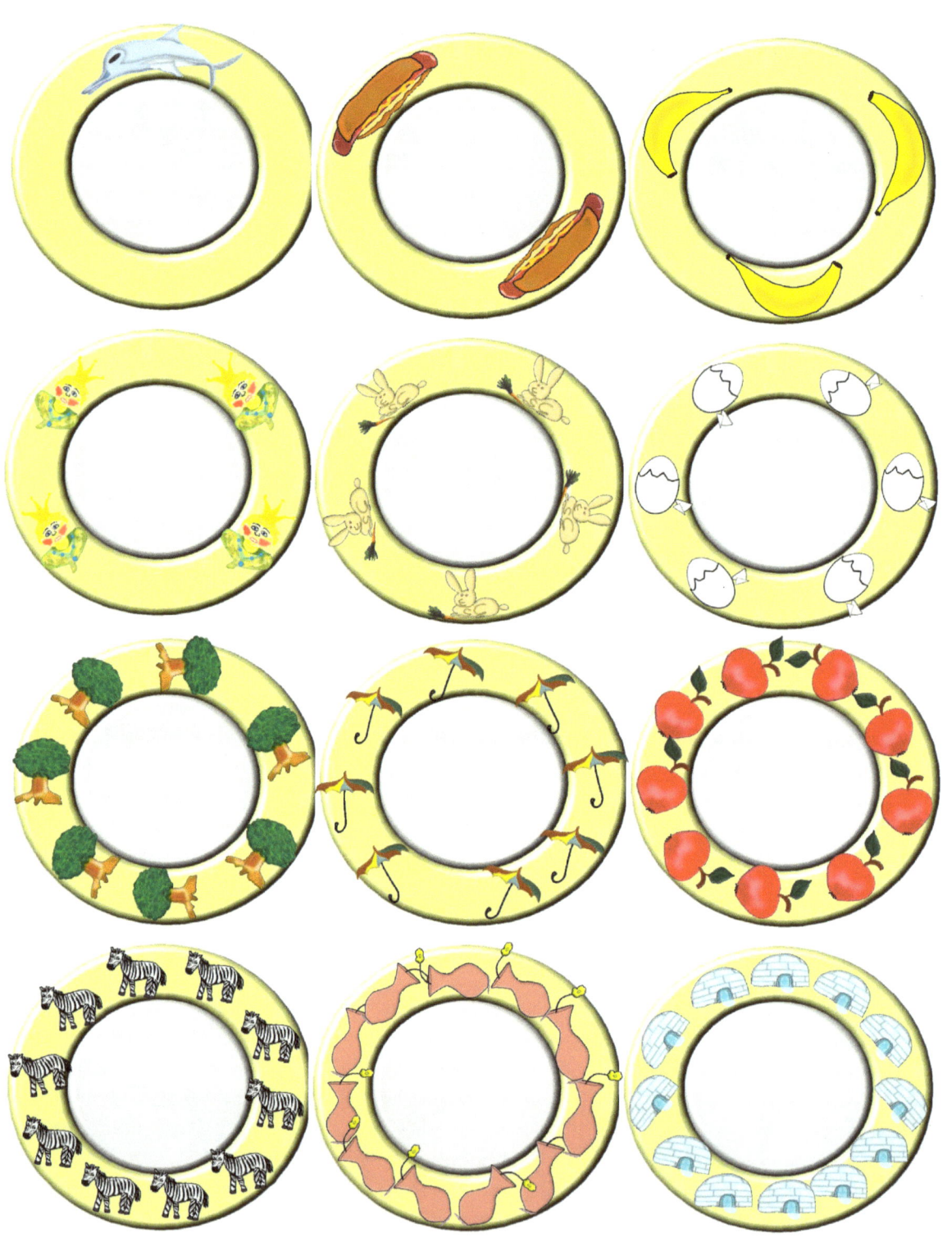

www.ingramcontent.com/pod-product-compliance
Lightning Source LLC
Chambersburg PA
CBHW051952210526
45473CB00023B/965